The O'Connor Line

Gen. Matt. O'Connor

John O'Connor of Navan m. Dorothy

George O'Connor m. Frances Berry d.1812

Dorothy Frances m. Pentland

Rev. Dr. John O'Connor m. (1777) Martha Weld

Rev. George Matthew O'Connor m.(1805) Frances Rawlina Nickson
1779-1842 — 1783-1838

John O'Connor 1806-1863 m. Elizabeth O'Keefe d.1863

Rev. Lorenzo Nickson 1815-1839

Dr. George Matthew 1817-1893

William Ford

Matthew Richard d.1902

Martha m. Samuel Gamett

Elizabeth Frances Ford m. William Tisdall

Lorenzo Nickson 1839-?

CHARLES YELVERTON O'CONNOR 1843-1902 m. Susan Letitia Ness 1849-1942

Frances Cecilia 1846-1915

Letitia 1850-?

George 1855-?

Aileen 1874-1955 m. C.Y. Simpson

George Francis 1875-1952 m. Olive Manning

Letitia Kathleen 1877-1968

Eva Droughaia 1878-1972 m. (Sir) George Julius

Chas. Goring Yelverton 1879-1950

Roderick 1881-1917 Killed in action

Bridget 1884-1979 m. Sir E.A. Lee-Steere

Murtagh Yelverton Going 1887-1959 m. Marie Lee-Steere

Charles Churchill (died in infancy) 1908
Patrick
Roderick
Avany

Maurice Young 1896-1977
Robert Barry 1899-1991
John Bell 1900-1983
Frank Harold 1902-1965
Charles Gilbert 1905-1954

Annie Eva 1903-1954
Patricia Eva 1916-

John d.1911
James

Marjorie d.?
Nancy d.?
Frances d.1963
Margaret

Charles (Killed in action 1942)
(Sir) Ernest Henry b.1912
Roderick Yelverton (Killed in action 1942)

Muriel b.1914
Elizabeth d.1980
Margaret d.1991

Charles
Roderick Yelverton

The Yelverton Line

The Yelvertons of Easton Maudit, Northamptonshire
16th–18th century noble family
Created Earls of Sussex 1717

Sir William Yelverton
(settled in Ireland)
m.
(1) Tuite
(2) Jones

Joseph Yelverton

George Yelverton of Belle Isle
m.
Letitia Burke

Walter Yelverton of Cork

Francis Yelverton m. Elizabeth Barry
1705–1746

Barry Yelverton m. Mary Nugent
1st Viscount Avonmore   d. 1802
1736–1805

William Charles Yelverton
2nd Viscount Avonmore
1762–1814

Barry John Yelverton
3rd Viscount Avonmore
1790–1870
m.
(1) Jane Booth
d. 1824

(2) Cecilia O'Keefe
d. 1876

William Charles Yelverton
4th Viscount Avonmore
1824–1883
m.
(1) Maria Teresa Longworth
(Annulled in Hse of Lords)

(2) M? Emily Forbes

Elizabeth Yelverton m. O'Keefe

Letitia Yelverton m. Charles O'Keefe

Sydney Eloise O'Keefe
m.
Lt. Col. Charles Foster Goring
(N.Z.)

Elizabeth O'Keefe m. John O'Connor
d. 1863   d. 1863

CHARLES YELVERTON O'CONNOR
1843–1902

C. Y. O'CONNOR

C. Y. O'Connor

His Life and Legacy

A. G. Evans

UNIVERSITY OF WESTERN AUSTRALIA PRESS

First published in 2001 by
University of Western Australia Press
Crawley, Western Australia 6009
www.uwapress.uwa.edu.au

Reprinted 2002, 2003, 2014

 The State of Western Australia has made an investment in this project
through ArtsWA in association with the Lotteries Commission.

Publication of this work was also assisted by generous funding from

 The Institution of Engineers, Australia—
Western Australia Division

 National Trust of Australia (WA)

 Fremantle Port Authority

National Library of Australia
Cataloguing-in-Publication entry:

Evans, A. G. (Anthony G.).
        C. Y. O'Connor: his life and legacy.

        Bibliography: p. 273–277.
        Includes index.
        ISBN 1 876268 62 X.

        ISBN 1 876268 77 8 (pbk).

        ISBN 978 1 87626 877 0 (pbk).

1. O'Connor, C. Y. (Charles Yelverton), 1843–1902. 2. Civil
engineers—Western Australia—Biography. 3. Water-supply
engineering—Western Australia—Eastern Goldfields. I. Title.

624.092

Lines from W. H. Auden's 'The Musée de Beaux Arts' cited on page 271 courtesy Faber and Faber Ltd
(*Collected Shorter Poems 1927–1957*, London, 1969).
Genealogical charts (endpapers) by A. G. Evans

Produced by Benchmark Publications, Melbourne
Consultant editor Amanda Curtin, Curtin Communications, Perth
Designed by Sandra Nobes, Tou-Can Design Pty Ltd, Melbourne
Typeset in 11pt Adobe Garamond by Lasertype, Perth
Printed by BPA Print Group, Melbourne
Reprinted in 2014 by Lightning Source

To our civil engineers,
whose work is indispensable and yet whose names are
seldom known to the public.

TO DONIE,

ENJOY OVER A GLASS
OF BRANDY,

TREV & MIKKI
X X X

Contents

Contents

List of Illustrations

Conversion Table

REFERENCES TO WEIGHTS and measures are given in the way in which they were expressed at the time, in imperial units. Conversions to the metric system are as follows:

1 acre	0.405 hectare
1 square mile	2.59 square kilometres
1 inch	25.4 millimetres
1 foot	30.5 centimetres
1 yard	0.914 metre
1 mile	1.61 kilometres
1 ounce	28.3 grams
1 pound	454 grams
1 ton	1.02 tonnes
1 pint	568 millilitres
1 gallon	4.55 litres
1 pound per square inch	6.89 kilopascals
degrees Fahrenheit (°F)	$\frac{5}{9}$ (°F−32) degrees Celsius

CURRENCY

Australian currency changed from pounds, shillings and pence to dollars and cents in 1966. Because of variations in currency values over time, actual conversions are difficult. At the time of the currency changeover, the following conversions applied:

1 penny (1d)	1 cent
1 shilling (1s)	10 cents
1 pound (£1)	2 dollars
1 guinea	2 dollars and 10 cents

Acknowledgments

THE ROMANTIC VIEW of the fiction writer closeted in a garret or a country cottage, producing a masterpiece isolated from human intercourse and alone, is far from the case in writing historical biography. In such a work as this, the author relies to a very large extent on the generosity, patience and cooperation of a whole army of advisers both professional and non-professional in various associated fields, and their help is acknowledged with gratitude. Without that generous help, this biography could not have been written.

In the early section on Ireland, I thank historian Danny Cusack, who allowed me to use his own valuable research material and introduced me to important contacts and sources of information in Ireland. In that country I was greatly assisted by Larry Duffy, the present owner of 'Gravelmount House'; Dr Pat McCarthy of Dublin; Patrick McNamee of 'Liscarton House'; Andy Bennett of the Meath County Library; local historians Oliver Ward of Nobber and Diana Pollock of Mountainstown; Julian Walton of University College, Cork; and staff of the National Library and the Public Records Office in Dublin.

In New Zealand, I have to thank genealogical librarians Richard Greenaway and Enid Ellis of the Canterbury Public Library Service; the most helpful and welcoming staff of the invaluable West Coast Historical Museum in Hokitika; and the Regional Archivist, Geraldine Pickles, and her helpful staff at the Christchurch branch of the New Zealand National Archives.

In Western Australia, the first person in a long list must be Frederick (Dee) Shelley, one time State manager of Mephan Ferguson, who lent me valuable research material; and then, in alphabetical order, Professor Geoffrey Bolton, who read the manuscript and made many helpful suggestions and corrections; Professor Frank Crowley; Alison Gregg of the Fremantle Local History Library; Richard Hartley; Dr Barbara Hewson-Bower; Bob Hillman; Tom Hungerford; Pat Hutchings; Harold Hunt; Ken Kelsall; Julie Lewis; Dr Toby Manford; Michael Price, archivist in charge of Private Archives in the Battye Library; Professor Martyn Webb; Stratton Yelverton; and Sharon Yelverton.

I am also grateful for the encouragement of the staff of University of Western Australia Press, particularly director and historian Dr Jenny Gregory, and to my editor, Amanda Curtin, whose corrections and suggestions were especially valuable. I am grateful, too, to ArtsWA for a Literature Grant, which helped towards the cost of travel to O'Connor country in Ireland and New Zealand. Without that financial assistance, the book certainly could not have been completed.

Thanks are due, too, to members of the O'Connor family for their encouragement and cooperation—to Mrs Muriel Dawkins and Judge V. J. O'Connor, who so generously lent me material from the family archives, and also Sir Ernest Lee-Steere, Mrs Pat Nuttall in London, and Dr Stewart Dunlop in Ireland.

Finally, I should not forget my patient family: my New Zealand travelling companion Antony Chesterman, who also read the manuscript and suggested improvements; my daughters, Alice and Emily, the former who contributed material help, and the latter for her wearying hours of important research in the Battye Library; and finally my wife, Claire, who read each section of the manuscript and, by her sensible comments, saved me from embarrassment.

A. G. Evans
August 2001

'You've never considered what it is that I do, have you? What engineers do?'

She looked at him steadily. Her face is expressionless and the dark masks her eyes. 'Tell me what it is you do, Will.'

He takes a deep breath and speaks very slowly. 'We change the order of things. And that is as dramatic as life gets.'

ROBERT DREWE, *THE DROWNER*

Introduction

FROM ITS POSITION high up on a granite plinth, set on a traffic island in front of the modern Fremantle Port Authority building, a dramatic bronze of Charles Yelverton O'Connor gazes enigmatically north-eastward over the great harbour that he designed and constructed—arguably his finest work.

And doubtless it would be of some satisfaction for that meticulous engineer to observe that apart from the modern cargo-handling equipment, which would seem strange to him, his essential design—its proportions, orientation, depth and wharf space—has not required much alteration since he completed it over a century ago. Like other far-sighted engineers of the late Victorian period with whom he shares a professional relationship, O'Connor built not only for the limited needs of his time but very much for the future: for our time.

The statue, like O'Connor himself, is larger than life. Its height (11½ feet) is difficult to gauge from the ground, but an average man standing beside it on the same plinth might reach as high as the figure's waist. The figure, modelled by Italian migrant Pietro Porcelli, is bare-headed, with one foot placed forward purposefully. The arms are folded, the right forearm raised towards the chin, suggestive of deep introspection. The left hand holds a roll of plans. The head from the front view is handsome and strong and only from the side reveals the curious rhomboid character of the crown, indicative, perhaps, of a difficult birth. It is said that O'Connor had to have his large felt hats—so characteristic of several of the pictures we have of him—especially made and imported from London.

The deep-set eyes stare out, and whether or not it was the artist's intention or perhaps his intuition, the facial expression is disturbing, having, like some ancient Etruscan bronze, a look accusing succeeding generations of indifference and ingratitude. Sir John Forrest put it more simply at a ceremony in 1912 when the statue was handed over to the Fremantle Harbour Trust, declaring that it was 'thinking in bronze'. He also spoke of O'Connor's integrity and honour, a man of unblemished character and untarnished reputation. Here, Forrest was attempting to lay to rest, finally, the vicious personal attacks made upon the engineer in the last months of his life, which undoubtedly contributed to his decision to end it. Forrest may also have been repairing his own feelings of

C. Y .O'Connor memorial by the Italian sculptor Pietro Porcelli. The Engineer-in-Chief looks north-eastward over his Fremantle Harbour from a high position outside the Port Authority building.

Photograph by the author

regret that he did not come to his friend's aid at a time when the engineer most needed him.

At the base of the plinth, on each of the four sides, are reminders of O'Connor's celebrated engineering achievements. Beaten copper plaques depict in turn Mundaring Weir, the Goldfields Pipeline Scheme, Fremantle Harbour, and—representing his work in planning and modernizing the State's railways— the Swan View Railway Tunnel through the Darling Range.

The subject of Porcelli's statue is in one sense unusual: a memorial to a civil engineer. Memorials of this grandeur are more likely to be erected to honour kings and queens, statesmen, generals and creative artists. Civil engineers are, more often than not, anonymous, and if considered at all are categorized as admittedly clever people but carrying out the proposals and instructions of others—servants of the state or the corporation. There survives, even today, a perception of civil engineers as 'other ranks', blue-collar workers who get their hands dirty, a perception dating back to the genesis of the profession in the British Army of the early eighteenth century. How many Australian school students, for example, could name the civil engineer who designed Sydney Harbour Bridge? Or the chief engineer on the Snowy Mountains Hydro-electric Scheme? Which Scottish schoolboy could name the engineer who built the Forth Railway Bridge? Or English schoolboy the chief engineer on the Channel Tunnel?[1] And yet these for the most part anonymous men, in the words of Robert Drewe, 'changed the order of things',[2] and contributed surely as much to the advancement of modern life—its conveniences, safety, economic benefit and health—as their counterparts in medicine, education and politics.

C. Y. O'Connor is one of the few who has not suffered this professional anonymity in Australia and, to a lesser extent, in New Zealand—certainly not at all in Western Australia. His name is well known to practically all young students who, armed with notebooks and pencils, make the mandatory school visit to Fremantle Harbour or to the C. Y. O'Connor Museum at Mundaring close to where his dam is located in the hills above Perth. He has become a legendary figure in the history of the State—his works are revered and his memory honoured—although whether this general interest stems as much from a fascination with the events surrounding his tragic suicide as from a true appreciation of his work is at least debatable.

The anonymity that, as has been argued, is the general lot of contemporary civil engineers is not altogether true of O'Connor's nineteenth-century mentors.

The early 19th Century was the age of the polymath virtuoso engineer...In no other period of English history were so many

engineers in practice whose names are still recognised. It was a time
of confidence and optimism.[3]

Thomas Telford, builder of canals, roads and bridges and the first president of
the Institution of Civil Engineers, is still a national hero in Britain and his works
form a tourist attraction. So, too, is Isambard Kingdom Brunel, a dominant,
flamboyant figure whose railway engineering, bridges, tunnels and shipbuilding
have raised him to the status of folk hero. And there were many others,
including the two Stephensons, father and son, and the other Stevenson,
Robert, famous for building lighthouses and improving harbours.

Telford had been dead nine years when O'Connor was born, but the
young student engineer was linked to the great man through his teacher, John
Chaloner Smith, a pupil of G. W. Hemans who himself was a pupil of Telford.
O'Connor most probably met I. K. Brunel and conferred with him because this
elder statesman of civil engineering, a Churchillian figure resplendent in top hat
and sporting a fat cigar, was retained as consultant to Bagnell and Smith when
O'Connor was a young member of the staff engaged on building Irish railways
(see chapter 4).

O'Connor was heir to that unique family of great Victorian pioneer
engineers, the men who designed and built modern Britain and Ireland, whose
dams, bridges, railways with their tunnels and viaducts, harbours and numerous
public works are still in use in those countries today. These engineers truly
'changed the order of things' and, using all their knowledge and skill, backed
their courageous plans, often in the face of opposition from more conservative
opinion. Like them, O'Connor always displayed unerring professional self-
confidence, but unlike most of them he joined that enterprising band of
engineer-adventurers who practised their skills in the new colonies overseas.
There they often had to contend with similar conservative opinion, and
occasionally—as in O'Connor's case—even with antagonism.

Pietro Porcelli's O'Connor,[4] derived from photographs, is indisputably
Victorian in character. Even more so than the long frockcoat, cravat and the
trimmed moustache, the stance and the way O'Connor holds his head exude that
confidence and optimism that are a mark of the late Victorian age. O'Connor,
like most members of his class, inherited a national confidence, a belief that the
social reforms of the period, together with the great advances in technology—the
second phase of the Industrial Revolution—would be the means of eliminating
poverty and ignorance and would eventually bring harmony and prosperity to all.

Nowhere was this pride, faith in progress and confidence in Britain's
achievements more evident than in the Great Exhibition of 1851. O'Connor, in

his ninth year, was probably one of the six million visitors (see chapter 3). The enormous locomotives, rotating machines and steam engines on show in Paxton's 'temple of glass', the Crystal Palace, were an awe-inspiring sight. Charlotte Brontë visited the Crystal Palace three times and wrote after her second visit:

> It is a wonderful place—vast—strange, new and impossible to describe. Its grandeur does not consist in one thing but in the unique assemblage of all things—Whatever human industry has created— you find it there—from the great compartments filled with railway engines and boilers, with the mill machinery in full work—with splendid carriages of all kinds, with harnesses of every description…It seems as if magic only could have gathered this mass of wealth from all the ends of the earth—as if none but supernatural hands could have arranged it thus. Amongst the thirty thousand souls that peopled it that day I was there, not one loud noise was to be heard— not one irregular movement seen—the living tide rolls on quietly— with a deep hum like the sea heard from a distance.[5]

From another contemporary account we learn that to stand in the machine room and be deafened by the pandemonic sound of hissing steam and clanging metal was a deeply moving experience overwhelming the minds of all with the thought of its possibilities. 'Many of the strongest were affected to tears.'[6] The Great Exhibition was an inspiration to all—particularly, one suspects, to young schoolboys fascinated by working models and great machines, leading undoubtedly to the kind of youthful ambitions now inspired by jet aeroplanes, computers and space shuttle technology.

Our contemporary assessment of the Victorian age is greatly coloured by contrasting the social services, public health and standard of living that we enjoy today with the harrowing accounts of poverty, neglect and child abuse portrayed in the novels of Dickens, Charles Reade and others: the sweat shops, the slums and the depiction of children chained to trucks pulling loads of coal in the mines. Such graphic evidence of the darker side of Victorian life is overwhelming to modern sensibilities. Historians generally agree, however, that by the middle of the century there was a marked improvement in general living standards as a result of a new reforming zeal. As tentative as these reforms may seem to us today, they were revolutionary in their time: the freeing up of trade, the repeal of the Corn Laws in 1846 (which removed a tax on the import and export of grain), the Factories Act of 1847 (limiting hours of work), the Public

Health Act of 1848 and the Poor Removal Act of 1846. All these, together with the economic and social benefits flowing from easier and cheaper transport, and a stability in the cost of living, contributed strongly to the extraordinary optimism of the period.

We all tend to judge our prosperity by contrasting the present with what is known of the past and not by some ideal in the future. The average Victorian, in whatever station of life, would have done the same. The reforms brought about in the twenty years prior to the Great Exhibition would therefore have been a cause for much optimism and pride. *The Times* of 16 December 1861, in a eulogy following the death of the Prince Consort, described 'the happy state of our internal polity and a degree of general contentment to which neither we nor any other nation we know of ever attained before'. Lest *The Times*'s opinion be dismissed as applying solely to its readers among the ruling and industrial elite, it should be noted that similar judgments were fairly common. The great Whig historian Macaulay, for example, wrote that 1851 'will long be remembered as a singularly happy year of peace, plenty, good feeling, innocent pleasure, national glory of the best and purest sort'.[7] The leading and much respected historian of the Victorian age, G. M. Young, wrote in 1936, 'Of all decades in our history, a wise man would choose the eighteen-fifties to be young in'.[8] Even twenty years before, the poet Coleridge was so moved as to write, when a reviewing a book devoted to the advances in 'Public Improvements', that he closed it 'under the strongest impression of awe and admiration akin to wonder. We live, I exclaimed, under the dynasty of understanding: and this is its golden age'.[9]

The questions may rightly be asked: to what extent, if any, could this new-found prosperity and optimism in England be shared by the O'Connors in Ireland, and was not the condition of Ireland, so recently devastated by the Great Famine and resultant mass migration, an altogether different case? As one observant foreign traveller stated, 'Ireland was the only part of the United Kingdom in which [was] found rural poverty and backwardness comparable with that to be seen in many parts of Germany, Switzerland, France, Spain and Italy'.[10]

To answer these, we should remember that Charles's father, John O'Connor, although he lost much of his money and his estate as a result of the Famine, was to some extent cushioned from its worst effects. He was never to lose his pride nor his place in the ruling Protestant ascendancy. He continued to be a magistrate. And his well-connected friends and his link with the aristocracy ensured that a suitable position was found for him when he had to relinquish his estate in County Meath. It can be argued, too, that young Charles, a small child at the time of the Famine, would still have imbibed from his class the self-same momentum, sense of civic responsibility, optimism and moral certitude of the Victorian age. As much as his own exceptional talents ensured that he would have a successful profession, so too did his family background and aristocratic connections. It may not be properly understood in the present day, when Ireland is recognized as a separate, proud nation, that in the mid-nineteenth century the Irish Protestant ascendancy would have considered themselves part of British life—Victorians—and proud of the empire. What was celebrated in England across the Irish Sea was also celebrated in the grand houses and on the wealthy estates in Ireland.

It is this poise of the quintessential Victorian that is so evident in Porcelli's statue and in all the photographs we have of Charles Yelverton O'Connor. He was, fundamentally, a man of his age, and if we are to get close to him and understand his character we must also understand the aspirations of the age— an age of intellectual upheaval, of enormous social and economic progress, and, in general, of moral and spiritual robustness arising from strong evangelical influences. Those ideals and values shaped O'Connor's life and eventually even the manner of his death.

His work may have transcended the age, but his character did not.

With this in mind I have attempted, in this new biography, to place O'Connor and his times in sharp focus against the background of his work. Previous biographies and monographs on O'Connor have tended to concentrate more on the political, economic and engineering detail derived from the wealth of government papers relating to his work. The result has been that O'Connor's work has become better known than the man. There is some excuse for this: while official government records remain in abundance, by comparison there are few of O'Connor's personal papers and letters in the archives. O'Connor himself was meticulous in keeping copies of his official papers and correspondence, as if this self-confident, intensely proud, professional man wanted the truth to become known to succeeding generations. Thus records exist relating to his professional decisions, and the less than fair treatment he received from the New Zealand and Western Australian Governments of the time. But there are very few personal letters revealing his own feelings. Doubtless O'Connor would have kept private letters and diaries in the same way that he kept his official papers, but, according to unverifiable family lore, many of these were destroyed soon after his death by his eldest daughter, Aileen, on discovering that she had been born before her parents' marriage. It is a tragic loss for posterity and particularly frustrating for those who attempt O'Connor's biography.

The story of the private O'Connor cannot be separated from the story of his public works; his works are the very reason for our interest in him as a man.

However, this biography does not set out to be a technical manual, nor a political history, although there are such works listed in the Bibliography for those who require them. In the following pages I have tried to shift the focus slightly and, while not neglecting the most important technical, engineering and political detail, to simplify the issues for the general reader and present a union of the man and his work with his times, each coexisting and inter-dependent upon the other.

How successful I have been in this approach to C. Y. O'Connor and his great engineering work must be left to the reader to judge.

PART I
IRELAND, 1843–1865

I

—

Lying north-west of Dublin, the ancient county of Meath is divided roughly in two by the historic rivers the Boyne and its tributary, the Blackwater. South of the rivers the land is flat plain, but the north is more varied, with hills and swells; the whole is verdant and fertile. When writing before the Famine in the 1840s, William Wilde, that assiduous traveller, described Meath as the great grazing ground of Ireland. He observed the crops as being so generally luxuriant, and the land so fertile, that had it all been sown with corn it would feed and might form the granary of the whole of Ireland.[1]

From an elevated position some 7½ miles north of the Boyne, down a little road heading south from the village of Castletown, which in turn is about two-thirds of the distance between the ancient town of Navan and the village of Nobber, stands the elegant country house known to this day as 'Gravelmount'.

Built, as its name implies, on a subsoil of gravel, it dates from the last years of the eighteenth century and is of characteristic late Georgian style. The ground floor is dominated by two tall sash windows either side of a central door crowned with a delicate fanlight and an entablature supported by Doric columns. The entrance opens onto a large hall, either side of which are the main reception rooms. The visitor, on entering the well-proportioned room on the left, is confronted by the original ornate, gilt-framed mirror on the far wall: a remnant far too imposing and heavy to have been removed by succeeding owners.

A central oak staircase leads up to the first-floor reception rooms lit by five similar-sized sash windows: clear evidence of a stately social life lived there a century or more ago. The bedrooms, all on the top floor, are lit by a smaller row of windows under the shallow-hipped slate roof, which is without eaves. Two chimney stacks rise up from either side wing. At the back of the house, and

'Gravelmount', the grand estate house at Castletown, County Meath,
leased by the O'Connors for fourteen years and where Charles Yelverton O'Connor
was born on the eve of the Great Famine.

Photograph by the author

in front of ancient farm buildings, a kitchen and scullery wing have been added. Below are extensive brick-vaulted cellars and semi-basement rooms for the servants. The same eighteenth-century architect built the neighbouring parish rectory, similar in design and scale, but this, sadly, was demolished by a speculator in the 1970s.

Today 'Gravelmount' is thickly covered from ground to roof line in a wall of green ivy and is being refurbished by the present owner as a country guesthouse. But in the middle of the nineteenth century, 'Gravelmount' would have looked youthful in bare honey-coloured local stone and would have been referred to by the locals as 'the big house'—a fairly grand estate house built, like many others, for the upper middle-class Irish Protestant families of the period.

The family claiming our particular interest is that of the O'Connors, who occupied the house and adjoining farm of 177 acres for a period of approximately fourteen years, dating from 1836. At that time, 'Gravelmount' was held on a long lease from Thomas Longfield, a prominent Irish judge and wealthy landowner. The agreements of those days, in practical terms, amounted almost to ownership, and would have allowed the O'Connors to alter and run the establishment in whatever way they pleased. As an example, the additional wing at the back of the house, greatly enhancing its facilities and value, was almost certainly added by the O'Connors.[2]

John O'Connor did not rely wholly on his income from the 'Gravelmount' farm but as much from rents derived from his inherited estate, 'Ardlonan', in the neighbouring parish of Kilbeg. There he owned some 340 acres, which were let to a number of tenants paying £2 per acre; in total, he would have received nearly £700 per annum—a large sum in 1840.

The first son born to the O'Connors in 1835 (and therefore not at 'Gravelmount House') was baptised George. In later years, he entered the army and served with the 9th Battalion of the Royal Artillery in India. George retired with the rank of lieutenant-colonel and inherited 'Ardlonan'. The second son, baptised Lorenzo Nickson, was born four years later and he, too, joined the army, serving with the Waterford Artillery in the American Civil War (where he was thought to have been killed).

In one of the bedrooms on the top floor of 'Gravelmount', a third son was born to John and Elizabeth O'Connor on 11 January 1843. He was baptised Charles Yelverton, the ceremony being performed in the local parish church of Castletown Kilpatrick on 28 February by the Reverend Edward Nixon. The father's trade or profession was given in the registry as 'Gentleman'.

The O'Connor family dated back to the seventeenth century and beyond. The Yelverton in Charles's name perpetuated a close link with the aristocratic Yelvertons of Northampton, one of whom was created Chief Baron of the Exchequer in 1795 and, in 1800, Viscount Avonmore of Derry Island. Charles's strong relationship with the Yelvertons was through both his mother's and father's side of the family. Two brothers, Joseph and Walter Yelverton, descendants of the old Northampton family, settled in Ireland in the mid-eighteenth century—Joseph in Belle Isle, Tipperary, and Walter in Cork. Charles's grandmother, Letitia, was directly descended from the Tipperary family; and his grandfather, Charles O'Keefe, was descended, through his mother, from the Cork Yelvertons. Adding further to the connection, and not a little to the reader's confusion, Charles's aunt (his mother's sister Cecilia O'Keefe) married a Cork Yelverton, Barry John, 3rd Viscount Avonmore. Having an uncle a viscount was to prove a valuable asset to the young Charles setting out on his career.

Charles's sense of compassion throughout his life and his firm evangelically based morality seem to have had their roots in both the O'Connor and the Yelverton sides of the family. His grandfather was the Reverend George Matthew O'Connor, rector of the parishes of Castleknock, Clonsilla and

Mulhuddart, County Dublin. He had married Frances Rawlina Nickson and died the year before Charles was born; his great grandfather, Reverend Dr John O'Connor, who married Martha Weld in 1777,[3] was rector of the same parish before him. They were known as gentle, scholarly men, both prosperous and owning country estates. The Reverend George O'Connor held the position of Moderator of Trinity College. At his death, 'Ardlonan' passed to John O'Connor. The Reverend George's parishioners at Castleknock erected a memorial plaque in the church in his honour, to which 'men of all religious persuasions subscribed'[4]—not surprising since the O'Connors had been rectors of Castleknock for more than eighty years consecutively. Charles's great grandfather had built the rectory, 'Hybla', which became the family home and figured largely in family history even into the twentieth century.

The Yelvertons, Irish by adoption, were public-spirited, compassionate men. The first baron, a judge, was described as 'one of the most competent and merciful men on the Irish Bench', and 'a great lawyer, an admirable speaker, and a statesman of sound and moderate judgement…and of a singularly sweet, simple, childlike nature. His only fault being irritability'. After his death, it was also charmingly written of him that 'he should on his death have had no more selfish wish than that justice could be administered to him, in the world to come, with the same spirit with which he distributed it in this'.[5]

Another account is less flattering. Yelverton's title was one of several created and bestowed upon members of the Irish Parliament in a none too subtle effort to bolster support for the Act of Union of 1801. He certainly changed his opinions but his eventual support of the union may have arisen from a genuine belief, held by Yelverton as by most other Irish politicians of his generation, that union with England was the only way forward for the Irish nation. However, the change of heart, whether for the best of reasons or not, did not protect him from attack by anti-unionists such as Sir Jonah Barrington, who wrote in *The Rise and Fall of the Irish Nation*:

> This distinguished man at this critical period of Ireland's emancipation burst forth as a meteor in the Irish Senate [but] after having with zeal and sincerity laboured to attain independence for his country in 1782, he became one of its Sale-Masters in 1800. His rising sun was brilliant, his meridian cloudy, his setting obscure.[6]

Keeping up the family tradition, Charles's uncle Barry John, the 3rd Viscount, was trained in law and became one of the registrars in the Court of Chancery. He inherited his grandfather's qualities of justice, kindness and selflessness. But not so his son, William Charles, the 4th Viscount and Charles's first cousin, who turned out to be the black sheep of the family and whose fall from grace will be told at the proper time (see chapter 4).

The O'Connors always remained proud of their ancient lineage, and even today some of the descendants include the name of Yelverton at baptism. The barony of Avonmore, named after a river in Cork and not the more famous river in Wicklow, lapsed at the turn of the century.

For an upper middle-class family at 'Gravelmount', members of the Anglo-Irish Protestant ascendancy, those early years at 'the big house' before the Famine would have been an agreeable, even halcyon period delineated only by the changing seasons on the estate and the lively social life of the district.

Judged superficially, there seemed to exist at that time a rare period of political stability. Ireland was a relatively tranquil place; there was no civil war nor invading army, and civil order was never threatened. But beneath the surface calm, the old hostility and sense of injustice long felt by the majority of the Irish poor towards England—a country widely considered an occupying power— were smouldering away ominously. Bitter opposition to the 1801 Act of Union was continuing, fanned by Daniel O'Connell's series of mass protest meetings throughout the country. The most menacing of these—or so thought the British Government—was the quarter million strong assembly on that symbol of Irish nationalism, the Hill of Tara. At the time, baby Charles O'Connor was scarcely three months old. The event turned out to be entirely peaceful, but Sir Edward Sugden, Lord Chancellor of Ireland, was not convinced. He wrote in his report: 'The peaceable demeanour of the assembled multitudes is one of the most alarming symptoms'.[7]

The sacred Hill of Tara being 14 miles south of Castletown, O'Connell's meeting could hardly have been unknown to the occupants of 'Gravelmount'. But we have no knowledge of John O'Connor's views on the union, nor on Anglo-Irish politics in general. It is clear, however, from certain of Charles's letters written much later in life, that Western Australia's Engineer-in-Chief had a strong sense of social justice that he surely inherited from his father. His sympathies were invariably with the labourers engaged on his work: 'I have continuously acted in what I have believed to be the best interests of the labouring classes, and have already procured for them the eight hour day'.[8]

Notwithstanding the portents from the direction of the Hill of Tara, the tranquil life at 'Gravelmount' continued for a while longer, and the only cries heard would have come from the stables and possibly from the direction of the nursery. John O'Connor bred horses and raced them successfully at the neighbouring Mountainstown Coursing Club. His mares Brunette and Zoe were noted as champions in the sporting columns of the *Meath Herald*. He was known throughout the district as a benevolent landlord, and an efficient farmer who keenly followed advice in the agricultural columns of the same newspaper on the importance of drainage—then essential to the development and health of the soil. He was appointed a magistrate and took his turn on the bench at nearby Kells. But John O'Connor was far from being the bucolic squire of fiction, interested only in hunting, shooting and reading *Burke's Peerage*. He had been classically educated by his father, and had then entered Trinity College at the age of 16, graduating as Bachelor of Arts in 1831.

Mrs O'Connor, when not pregnant, was heavily involved in the social life of the district, and held balls and dinners in her newly furbished Georgian reception rooms on the first floor of 'Gravelmount'. The gravel forecourt illuminated by candelabra and oil lamps shining down from the elegant windows of the reception rooms must have been a pretty and welcoming sight to those in their carriages lucky enough to be on the O'Connors' guest list. Mrs John O'Connor's name was often noted in the social columns of the *Meath Herald*, and she was present at the annual Summer Shows of the Meath Agricultural Society. These popular events held in the Navan courthouse were fully reported in the newspaper and at least one was celebrated with a gushing poem. Prior to the Famine years, the family consisted of three boys, George, aged 10, Lorenzo, 6, and Charles, rising 3 years. Two children were to follow: Frances Cecilia, born in 1846, and Letitia, who came on the scene after the family had departed 'Gravelmount' for the south of Ireland.

The family tradition of ecclesial service with its attendant doctrinal teaching and liturgy would have been strongly reflected in family life at 'Gravelmount'. John O'Connor, born and brought up at 'Hybla', the elegant rectory near Castleknock, being the son and the grandson of devout if rather select clergymen (and the nephew of yet a third), would have insisted, as a matter of course, on strict religious observances by his family, including the servants. Whether the young O'Connors would have 'heard delivered or read aloud, a thousand sermons', as young Victorians were described as doing by historian G. M. Young, they certainly would have been required to attend morning and evening prayers and to read the Bible on Sundays after attending morning and evening church services. Young recounts how Robert Peel—the Prime Minister of that time—was trained by his father to repeat verbatim every Sunday the discourse he had just heard in church (which is said to have accounted for Peel's prodigious memory).[9] Charles certainly grew up with a well-developed memory and we have evidence that wherever he was, he would always attend church on a Sunday if at all possible.

The comfortable, gentle rural round at 'Gravelmount' was destined not to last. The first rumblings of the tragic potato crop failure were being heard in the autumn of the year 1845. And those rumblings soon grew into a terrible storm that shook the country, produced untold suffering, decimated the population, and left no one, neither landlord nor peasant, unaffected. It was a defining event in the history of Ireland whose effects are still remembered with some bitterness today.

And yet, unlike the proverbial ill wind that blows nobody any good, the Famine produced a chain of events for the O'Connors that was to the eventual benefit of New Zealand and Western Australia: without it, C. Y. O'Connor may never have set foot in either country as a civil engineer.

2

———

THE HUMBLE POTATO was the mainstay, and very often the only food, of nearly half of the Irish population. Based on the census of 1841, it would therefore follow that more than four million souls depended not on their daily bread, as in the Lord's Prayer, but on their daily potatoes, and that when the crop failed, as potatoes were wont to do without benefit of later scientific breeding, the poorest people faced starvation. 'The unreliability of the potato was an accepted fact in Ireland, ranking with the vagaries of the weather.'[1]

When the crop was healthy, as we may suppose it was more often than not, the potato provided a cheap, nutritious but pleasant diet. It was capable of being served in a variety of ways. A family of five or six could live healthily for twelve months on the produce of an acre and a half. But the possession of this amount of land by the poorest peasants was rare as a consequence of ever smaller subdivisions of land owned by both absentee and some resident landlords, and administered by their greedy agents. We know that five years prior to the Famine, John O'Connor's estate at 'Ardlonan', which he inherited from his father and grandfather, was nominally leased out in three parcels of land: to Patrick Reilly, 32 acres; to Henry Smith, 58 acres; and to Anne Reilly, 201 acres.[2] Although John was known to be a compassionate man, and cared for his tenants, it is probable that each of them would have sublet, and these lessors could have sublet yet again. Clauses against subdivision existed in some leases, but 'to put them into operation [was] dangerous'.[3] John O'Connor was also likely to have sublet part of his leased 'Gravelmount' estate. Subletting was not always motivated by profiteering: true, it was widely practised because it provided a regular income without investment, but some of the better landlords would, as a moral necessity, have let small parcels of land on conacre[4] to needy families. Unless a labourer could obtain such land for growing potatoes, his family would starve. Large farms were rare. The 1841 census shows that the subdivision of land had reached a point where just about half of all Irish landholdings were smaller than 5 acres.

The terrible blight, the fungus *Phytophthora infestans*, which was to have such calamitous consequences, made its first appearance in 1840 in North America and was later brought across the Atlantic to Europe, probably lurking in the garbage dumped overboard from a ship. The first indication that it had arrived in England was a letter to the Prime Minister, Sir Robert Peel, reporting that a crop had failed on the Isle of Wight in August of 1845. England was not immune from the blight, but the consequences were not as severe in a country where the staple diet was bread and cheese rather than potatoes. In comparing the two, Arthur Young wrote:

> I will not assert that potatoes are a better food than bread and cheese
> but I have no doubt of a bellyful of the one being much better than
> half a bellyful of the other.[5]

The alarm was not raised in Ireland until September, when the first outbreak was noticed in Waterford and Wexford. Even then, opinion differed about the seriousness of the disease; July had been an exceptionally dry and warm month and the year's crop was expected to be healthy and plentiful.

Perhaps the area around Castletown and 'Gravelmount' might be spared? The *Meath Herald* seemed to think so. The edition of 27 September described as 'gross exaggeration' ominous reports and warnings of disease in other parts of the country and advised its readers 'not to give credence to them'. However, in the edition of 11 October, the same newspaper was forced to make a humble apology admitting it had been wrong:

> We gave it as our opinion that the potato disease had not extended
> its baneful effects through the crops in this district—and at that
> period its visibility was not perceivable. Would that we could give the
> same opinion now, but alas, and it is with deep regret, we feel
> ourselves obliged to state that the disease has set in, and to the most
> alarming extent within the last ten days.

The *Herald* had some excuse for complacency because, at first, the worst hit areas were in the south and west of Ireland, but within a short time other parts of the country were also affected. Ironically, the newly harvested potatoes often showed no sign of blight. But this proved no cause for celebration. After only a minimum time in storage, they would be discovered to have degenerated into a black, slimy, decaying pulp. The outlook for the following year was terrifyingly simple: the small percentage of the crop not affected, and which in good years would have been kept back for seeding, would, naturally enough, be consumed immediately.[6] According to a government estimate, within four months of harvesting, the entire 1845 crop—between 9 and 10 million tons of potatoes—had been lost: the equivalent of one and a half million acres sown.[7]

The young Charles O'Connor, just 3 years old, could not have been aware of the disaster in the countryside around him, although his family was directly affected, not by a shortage of food—for their diet would have been nutritious and varied—but by the inability of tenants to pay rents, and by a distress felt for local people in a far worse situation than themselves. Even before the Famine set in, the Poor Inquiry Commissioners reported that the suburbs of Kells—the administrative centre nearest to 'Gravelmount'—contained 'the greatest misery which they had seen in any country'.[8]

Charles's father was one of those quick to involve himself in emergency welfare measures. Ten days after the *Herald*'s volte-face, a meeting of land-holders was held in the Castletown schoolhouse, chaired by the local vicar, Reverend Edward Nixon. Information given confirmed that more than half the potato crop in the parish was diseased and that the fungus was still spreading. The meeting resolved that John O'Connor, JP, should communicate these facts to government authorities and urge help in alleviating the coming distress.

A second meeting was held three weeks later with O'Connor in the chair. The purpose was to read O'Connor's report and also the brief and less than comforting acknowledgment from the office of the Lord Lieutenant in Dublin, expressing His Excellency's regret at the spread of the disease: 'The earnest attention of Government has been directed to the subject'.[9] At this second meeting, speakers unanimously agreed that there had been a 'visible increase' in the number of crops affected. Two of the speakers confirmed that they had visited several properties nearby and 'had scarcely found one sound potato'. Several landowners came forward 'determined to remit all claims for the potato rent'. A ten-point plan was agreed upon, which included the appointment of a committee to establish a fund to buy meal and sell it at the lowest possible price, and with the power 'to distribute gratuitously to such as they shall find in absolute want'.[10]

Further evidence of John O'Connor's profound distress at the condition of his tenants, together with others affected in his district, is provided by an official report to the government-sponsored Relief Commission by one of its agents, Charles S. Clements. After quoting O'Connor's conclusion that on average two-thirds of the potato crop had already gone in his district, Clements states that O'Connor, along with eight other local landholders, had agreed to remit the rents of his tenants, and, further, that John O'Connor had 'afforded the best proof of his sincerity by having within the last week or ten days taken twenty additional hands into his employment, fifteen being his usual number'.[11] No need to wonder that, with gestures like these, John O'Connor could not continue to run a profitable estate if the Famine continued for any length of time.

And continue it did, with disastrous results. Within two years, the agricultural character of the county changed: the acreage under potatoes declined from 48,245 in 1845 to 4,573 in 1847, a reduction of 90 per cent.[12] Over a period of four years, twenty thousand deaths occurred in Meath alone directly as a result of the Famine, and this, together with the number emigrating, resulted in a decline shown in the population census returns for Castletown from 3,011 in 1841 to 1,722 in 1851.[13]

In February 1846, the Relief Commission set up local committees throughout Ireland to collate information, administer aid, raise subscriptions, and make recommendations to the Board of Public Works for construction schemes to be used as a means of employment. Rules were published for the conduct of the committees, which also had the task of raising local collections from landholders. Money thus raised was matched by commission grants. When some landlords refused to contribute, the committees were instructed to publish their subscription lists, and the names of non-subscribing landlords were sent to Dublin Castle. (One of O'Connor's early remittances is noted as £60; doubtless there were more to follow.)

John O'Connor was elected to the Navan Central Relief Committee, which met weekly in the courthouse. Often the central committee elected sub-committees to cover distant parishes—for example, the one in Castletown was presided over by O'Connor himself. At first, committees were prohibited from giving food gratuitously unless it was established that there was no room for the needy in the local workhouse; in practice, this was often ignored and free food was distributed on the committee's own initiative.[14] Although many committees in various parts of Ireland attracted criticism, one government report found that the committees in Meath were 'acting independently and conscientiously', and a senior official reported to the treasury in December 1846 that 'the greater part of the Committees performed their duties most spiritedly, meeting day after day to help stave off the impending calamities'.[15]

One of O'Connor's prime concerns was the provision of additional employment in his district so that the destitute could buy grain; he himself supplied grain from 'Gravelmount' at lower than market prices. He argued strongly for an increase in public works and was a forceful speaker at a public meeting in Navan in February 1846, where he defended the building of a new road. Strong objections to the road were voiced by some of those present, partly on account of the expense falling on local cess (rate) payers, but more significantly because its alignment would encroach on the land of several of the landowners' properties. John O'Connor, however, accused them of denying work to a hundred unemployed, and in the course of his speech 'made some short but touching comments on such landlords as held aloof in a time of necessity—and why they should not aid in providing employment and relieving the poor'.[16]

By the middle of the first year of the Famine, there were 648 local relief committees throughout Ireland.

In addition to his relief committee work, John O'Connor was elected on 28 March 1846 to the Board of Guardians of the Navan Union, which administered the local workhouse. Each week, the board's report was published in the local newspaper: it included the number of persons in residence, the number admitted, and the number discharged or who had died. The board also set the diet for inmates and kept the accounts; in Navan's case, the cost of provisions for one week was given as £34 10s 6½d—an average expenditure on each pauper of 1s 5½d. For this amount, the inmates received an allowance of buttermilk daily, bread on three days a week, and soup on the other days. In the week under consideration (immediately prior to John O'Connor's election to the board), the inmates at Navan workhouse numbered 443; nine more were admitted within the week, eleven were discharged and three died. In some areas, the workhouse diet compared unfavourably with that served in local gaols and there are instances of inmates committing crimes to get themselves transferred to gaol.[17]

John O'Connor's constant round of committee meetings and relief work, and the business of the farm, must have left little time for leisure. It is not hard to equate his situation with that described so graphically by historian Thomas O'Neill in his study of the administration of famine relief:

> Many landowners, though greatly impoverished by the inability of their tenants to pay rents, assisted in every possible way and denied themselves all luxuries in the crisis. On them fell the burden of work on local committees, as well as the duty of subscribing to the funds of the local bodies. They were frequently poor law guardians and their large estates sometimes extended into several relief districts and multiplied calls on the owner's time and income.[18]

The work of the committees was additionally frustrated by the massive Relief Commission bureaucracy, which required the keeping of exhaustive records and the issuing of a constant flow of sometimes conflicting regulations governing the distribution of aid. As we have seen, in some parts of the county these provisions—for example, the regulations denying aid to any person owning even minimal livestock, however unhealthy, or owning more than one acre, however unproductive—were occasionally ignored.

By the winter of 1846–47, the food position had become desperate: widespread starvation reigned. In the Navan district, 'several bakers and provision shops were attacked by large crowds of persons crying out for bread… the assailing parties were dispersed without doing any injury'.[19] Sir Robert Peel announced in the House of Commons that 'the prospect for the potato crop this year is even more distressing than last year—its ravages are more extensive'.[20] On 15 October, the *Meath Herald* reported that John O'Connor had called an emergency meeting of the proprietors and occupiers of land in his neighbourhood and passed a resolution stating that because they

> were peculiarly circumstanced from the number of labouring people
> in relation to the acres and territorial particulars we feel the necessity
> of combining at once the energies of all parties therein interested.

To add to John O'Connor's worries, the already calamitous situation in the Navan district was worsened by an outbreak of a virulent strain of typhus. The fever had been prevalent in earlier years but was aggravated by additional squalor and poor nutrition, and by the movement of scores of destitute and starving wretches from the south-west of Ireland towards what they considered to be—with little basis in fact—the lesser affected areas of the more fertile north-east.

The Irish winter of 1846–47 was the most severe in living memory. Snow and frost were continuous and these unusual conditions added to the general misery and suffering of the people. Government soup kitchens were established, and by August 1847 over three million destitute were receiving food provided by them. The great migration had started: many sailed across the Irish Sea in search of work and food in Lancashire factories; others crossed the Atlantic to Canada and the United States. Amid the starvation and misery in Meath, a wealth of corn was being produced—as observed earlier by William Wilde— but this was mainly for export to Britain and elsewhere at high prices. Instances are recorded of carts transporting corn for export being guarded by soldiers against attack by desperate mobs. That John O'Connor, like a small number of other sympathetic landlords, did not dispose of corn in this way but chose to sell it much more cheaply to the starving poor not only contributed to his worsening financial situation but actually placed him and his farm workers in some danger. The greedy middle-men, the meal-mongers, put pressure on the landlords to sell them their harvest. In one instance in Meath, a corn producer, Connell, who was known for distributing his corn cheaply to the poor instead of selling at better prices to profiteering agents, was pulled off his cart when returning with his sister from Mass in Kilskyre, and beaten to death.[21]

At the beginning of the third year of the Famine, little Charles O'Connor turned 5 years old and by this age must have been slowly becoming aware of the situation in the countryside around 'Gravelmount'. He might have noticed the strain on the faces of his parents and listened to the stories told by his elder brothers. Doubtless he was sheltered from the worst horrors, as children of those times tended to be, but a dutiful, socially concerned father such as John O'Connor would have likely instilled in his son—especially at mealtimes—his good fortune when compared with the starving children in his immediate neighbourhood. That 'for the children these were happy and exciting days', as Tauman has related,[22] is open to serious doubt. If so, those days did not last long for young Charles.

The finances of the Board of Guardians were derived from local rates on properties. Landlords were still required to pay their rates, although in many cases, including John O'Connor's, rents for properties were of necessity waived during the Famine. The rateable value of O'Connor's 'Ardlonan' property alone was, in 1854, more than £370 per year and the landholder received no lenience from the tax collectors. 'The famine which swept away so many small farmers did little good for the landlords...many a landlord suffered much in the years of shortage and depression.'[23] Inevitably, as the Famine continued its course through its third and fourth successive years, the effect on 'Gravelmount' and on John O'Connor's ability to maintain the house, farm and estate was ruinous.

Exactly when he decided to give up is not clear. Like many another landlord, he would have been, if not technically bankrupt, in a practical sense nearly so. Rents would have been forfeited, rates unpaid and owing, and a general air of wretchedness and disease would have permeated the countryside. Landlords who owned thousands of derelict acres were reported virtual prisoners in their own mansions, without tenants or workers, having to exist on rabbits shot in their overgrown parks.

We have no written evidence of the O'Connor family's last months at 'Gravelmount', or the decision to leave and in exactly what circumstances. But we can imagine how careworn they must have been, and how regretful to leave their home and the intimacy of local society after fourteen years. A new 'Gravelmount' lease was registered in the name of Thomas Hopkins in 1854, but we know that John O'Connor had certainly left by the close of 1850. Whether his wife and some of his children remained in residence there while John was disencumbering himself from his lease and taking up employment in the south, we cannot be sure. He was then 44—an age considered more advanced in years in those days than it is today.

He arranged for Charles, then aged about 7, and his youngest child, Frances, to be sent to his sister, their Aunt Martha. She and her husband, Samuel Garnett, lived at 'Summerseat', Clonee, in the parish of Dunboyne on the borders of Dublin and County Meath. Again we do not know the exact date of the transfer, but presumably splitting the family in this way was a welcome option in the circumstances of the Famine itself and the crisis of moving and resettlement.

The Garnetts lived in a stately two-storey Georgian house slightly smaller than and of different proportions to 'Gravelmount'. Situated 10 miles from the capital, 'Summerseat' nestled in the midst of a tranquil, well-tended park, part of a wooded estate of 155 acres. The house is still there, hidden from the road

'Summerseat', in the south of Meath, near Dublin, where the young Charles and his sister Frances were sent to live with their uncle and aunt during the Famine.
Photograph by the author

and the neighbouring village by an avenue of beech trees. When the young Charles came to live at 'Summerseat', the estate was renowned throughout the district for its horses and the copious stables, which have since been sold and now form part of an adjoining residence. Samuel Garnett was master of the Meath Hunt. It has been recorded that C. Y. O'Connor's lifelong passion for horses and riding dates from his time at 'Summerseat'—although from an even earlier age he would have grown familiar with his father's stables at 'Gravelmount'.

In the local Dunboyne Church of Ireland, a white scroll on a black marble surround can be seen on the south wall, inscribed to the memory of W. S. Garnett and his wife, Marianne, the parents of the younger Samuel Garnett, who became John O'Connor's brother-in-law. Garnett the elder was a prosperous landlord who farmed a large estate at Williamstown neighbouring the O'Connors'. A fellow justice of the peace, his name featured prominently alongside John O'Connor's at local social events and on various committees. Evidently the two families were intimate friends and in this way John's sister Martha would have met and married the younger Samuel Garnett. At the time of Charles's residence at 'Summerseat', the elder Garnetts were still alive, the old man aged 75, and his wife, 52.

Charles's years at 'Summerseat' were evidently happy. There he learned to ride horses, and was tutored by his aunt.[24] A small Protestant school had been established by Squire Garnett in the village, and it is likely that Charles attended lessons there; perhaps his aunt was in charge. It may have been possible to forget the horror of the Famine at 'Summerseat': the misery was less severe in southern Meath, so close to Dublin, where wheat and cattle farming predominated. And if Charles and his sister were at 'Summerseat' by the first week in August 1849, as is probable, they may well have been taken to see the arrival of Queen Victoria and Prince Albert on the Royal Yacht as it sailed into Kingstown Harbour on the 9th. In the midst of the Famine, the royal visit was not expected to arouse an enthusiastic response from the Irish people. However, according to the normally unsympathetic *Freeman's Journal*, 'The personal demeanour, the frank and confiding manner of the Queen, have won her golden opinions'.[25] After a somewhat icy start to the tour, public opinion soon thawed and people came in their thousands to cheer and wave. The quays were crowded, and dense masses of people strained for a view of the royal processions. Banners and flags were flown from office buildings; roofs and windows were jammed with faces. It was the greatest and most colourful free show that the poor people of Dublin had ever witnessed. It is more than likely that Mr Samuel Garnett, JP, master of the Meath Hunt, landlord of 'Summerseat' and member of the Church of

Ireland, together with his wife, would have received invitations to one or more of the royal receptions, living as they did so close to Dublin.

But while the Garnett family and their young charges are enjoying a comfortable social life (possibly enlivened by glimpses of Her Majesty, her Consort and three of their six children), and well isolated from the full effects of the tragedy of the Famine, we should turn back to John O'Connor, who was about to leave 'Gravelmount' and embark on a new and wholly unfamiliar path—that of a salaried servant of the newly formed Waterford & Limerick Railway Company located in the far south of the country.

3

FOLLOWING THE SUCCESSFUL opening of the Stephensons' Liverpool–
Manchester Railway in 1830, 'railway mania', which gripped England during
the early years of Queen Victoria's reign, crossed the Irish Sea and similarly
transformed the fortunes of the few (for better or worse), and generally
improved the lifestyles of the entire population. With passenger fares at about a
penny a mile, for the first time the ordinary people of Ireland were able to travel
relatively cheaply beyond their immediate locality, and thousands of new, secure
jobs and careers were made available. The growth of countrywide railways was
so rapid throughout Ireland after the standardization of the gauge in 1847 that
more than 400 miles of line were in commission by 1850 and 840 miles by
1853, and in the subsequent twenty years of development the number of
passengers reached more than fourteen million annually.[1]

If development and investment in Irish railways faltered during the
Famine years, optimism and enterprise on the part of the railway entrepreneurs
did not. Glowing advertisements urged investment in new companies and new
lines; news of railway building projects and operations and regular company
reports filled newspaper columns. Even leading articles offered rhetoric, and
sometimes criticism, as if railway business was the primary concern of the
whole nation.

The first railway service, which had opened in December 1834, linked
Dublin and Kingston, but the first line actually sanctioned by parliament—and
then not brought into service until much later—was the Limerick and
Waterford Railway in 1826. Not until 1845 was the reverse-named Waterford
& Limerick Railway Company established, but its initial momentum was
restrained by the Famine. Money was scarce and a treasury loan was raised so
that the first section of the line, between Limerick and Tipperary, could be

completed—and that was largely achieved (not without objection) by the famine relief employment scheme.

Railway mania often led to several companies competing in proximity to one another: at one time, five such companies had connections with Waterford, and full amalgamation did not occur until 1924.[2] In those early days, railway companies often ran their trains over lines owned by a rival company and paid charges estimated at a few pence per train mile. By 1867, the Waterford & Limerick Railway Company owned 77 miles of track but worked its trains over 150 route miles; the cost of paying these fees was a constant restraint on profits.[3]

Although the company was registered in Waterford, the line commenced in Limerick. A critical leader in the *Waterford News* of 5 July 1850 complained of 'painfully slow progress' and demanded that construction work begin 'not in Ballinasloe, or Limerick, or Tipperary, but in Waterford'. It pointed out that the people of Waterford had contributed the most money towards the undertaking 'and they should not be the last to participate in its advantages'. On the same page was a verbatim report of the shareholders' meeting in the Waterford Town Hall three days earlier. Cost-cutting was the chief concern, and an important item on the agenda was a proposal to reduce by half the secretary's salary, which then stood at £600 per year. Evidence was given that other companies were paying their secretaries considerably less, and all agreed that economy was necessary in the difficult circumstances then being experienced. The debate elicited considerable reaction, not least from the secretary himself, Mr R. Saunders. He pointed out that he had been not only secretary but treasurer, traveller and financial manager of the company and therefore could not submit to a reduction, and if the motion were passed he would have to leave. There was much heated discussion for and against the motion. One major shareholder named Mortimer, with an eye to the practicalities, reminded the gathering that important negotiations were pending for August, and as the secretary was the principal witness in the case it would be impossible to proceed without him. Mr Saunders graciously solved this problem by saying that he was willing to work for as long as the salary remained at £600 per year, and only when it was reduced would he leave.

Finally, on an amendment, the motion setting the secretary's salary at £300, to take effect from 1 January the following year (1851), was carried unanimously. The Waterford & Limerick Railway Company had six months to find a replacement secretary.

How John O'Connor was appointed to the vacant position with the company—whether he responded to an advertisement or whether, as seems more likely, he was recommended by one of the directors whom he knew personally—

is not known, but he was listed as a shareholder and would doubtless have been familiar with the company's activities. We know that he took up his duties on 1 January 1851 at the stated salary of £300 and was responsible for calling his first shareholders' meeting of the year, held in the Waterford Town Hall, on 27 February. His address was still given as Castletown at that time, and he retained his title of justice of the peace. It is probable that he resided temporarily in Limerick, where the company's offices were then located at the railway station. His family did not join him in the first months of his new life.

We may wonder what qualifications the former farmer and estate manager John O'Connor brought to his new position. True, he was a university-educated man and his recent involvement on semi-government famine relief committees, which managed large public funds necessitating a good deal of administration, would have prepared him to some extent, but he could not have known much about company law and business procedures. As his predecessor complained— and John O'Connor was soon to learn—the role of the company secretary combined that of manager, treasurer, traveller, government department negotiator and general factotum. He would have had to handle all the business of the company on behalf of the board and shareholders: a heavy responsibility. A newspaper report shortly after his appointment declared that the management of the company 'had been vested in men utterly unacquainted with the merest elements of the routine of the transport business'.[4] This condemnation appears to have been directed more at the constantly changing board, whose chairman was the Reverend Mendicott, a gentle, ineffectual clergyman, but it cannot be discounted that the paper also had the new secretary of the company in its sights. However, judging by the same newspaper's report of O'Connor's first half-yearly shareholders' meeting in Waterford, one month after his appointment, he seems to have acquitted himself well. The directors' report, read by O'Connor, announced among other matters that instead of declaring a dividend for the half-year, profits would be used to liquidate the liabilities of the company. It stated that the construction of 52 miles of line to Waterford, still to be completed, was estimated to cost a little over £8,000 per mile. He also announced that a Bill to authorize certain deviations in the Waterford–Limerick line was then before parliament.

Railway legislation of the period required companies to obtain sanction for lines, and any subsequent deviation of lines, through an Act of Parliament. Proposals put forward by companies were first considered by a select committee before being presented to both Houses, and one of John O'Connor's duties would have been to prepare all the documentation, travel to London and appear before the committee as a representative of the company. Inevitably, O'Connor would have visited London on several occasions, and certainly would have done so in his first year, 1851. Once in London on railway business, what would have been more natural—even essential for his own and the company's benefit—than a visit, or several visits, to the Great Exhibition, which opened in Hyde Park on 1 May? The exhibition was not only the greatest spectacle and entertainment the general public had ever witnessed but served as a trade fair, where the latest advances in industrial machinery and production attracted those professionally interested. Railway engines, carriages and the latest railway equipment were displayed, and among the exhibits were the first railway brake system calculated to prevent collision, and the hydraulic buffers that became such a characteristic feature of all subsequent steam locomotives.

And if John O'Connor visited the exhibition, is it too fanciful to think that he would have brought with him his young son Charles, rising 9 years old? And even if John did not bring him, would not Charles's guardian, Samuel Garnett, have done so?

The Great Exhibition attracted more than six million people over seven months: visitors streamed in from all over the country, special reduced fares and excursions were promoted, and charitable and church organizations subsidized visits for poor children so that no one was excluded.

> Agricultural labourers under the guidance of their clergymen arrived
> in London…wearing their peasants attire, and an old woman of
> eighty-five walked the whole way from Penzance with a basket on her
> head for the express purpose of seeing the exhibition.[5]

To visit the Great Exhibition was the goal of nearly everyone in the kingdom from the Queen and the exhibition's progenitor, Prince Albert, down to the humblest subjects. Even comparison between the overwhelming popularity of the Great Exhibition of 1851 and, say, the World Cup or the Olympic Games of today would be unsatisfactory: in the case of the contemporary examples, there exist many competing distractions and amusements, but the Great Exhibition was unique and the event of a lifetime.

At the half-yearly meeting of the Waterford & Limerick Railway Company on 16 June 1851, John O'Connor read the summons convening the meeting and the Bill to authorize certain deviations, which the shareholders were asked to approve. The meeting was told that the Bill had passed the committee stage and the House of Commons, and was then before the Lords.

The business of the company proceeded, the railway expanded, and regular half-yearly company meetings from this date forward were reported in the local newspapers, with John O'Connor well in evidence as executive. Some time before 1857, the main offices of the company were transferred to Waterford City, to the satisfaction of the townspeople and surely of the O'Connors, who were by then united and living in the vicinity.

Life in or close to the city of Waterford must have provided a striking cultural shift from farming in rural Meath, where the closest town to 'Gravelmount' was Navan. In comparison with Waterford, Navan was a backwater, 'a dirty, ill-built, straggling collection of houses, boasting the honour of having been half a country-town'.[6] The O'Connors' new home was the chief town and commercial port of the county that bears its name, situated on the river Suir at the head of a tidal estuary. At that time, Waterford had a population of around 20,000 and was second in importance to Cork, its near neighbour. The city, 111 miles south-south-west of Dublin, boasted two cathedrals, a fine town hall, a customs house and barracks, and the eleventh-century Reginald Tower famous for one of Cromwell's cannon balls being lodged in the stonework. Waterford's prosperity depended on its breweries, distilleries and flour mills, deep-sea fishing, and the export of cattle, sheep and agricultural products. From the bustling quayside, focal point of the town, ships sailed for most Irish and English ports and the Continent. From the same quay, John O'Connor would have boarded one of the frequent packet-boats that plied between Waterford and Bristol. There he could connect with Brunel's renowned modern train service, which thundered across towering viaducts and elegant bridges, and through one of that engineer's most remarkable triumphs, the 1⅘ mile Box Hill Tunnel,[7] completing his journey to Paddington Station, London, in a little under four hours.

It was these recent marvels of the Industrial Revolution and John O'Connor's association with his engineer colleagues—especially G. W. Hemans, pupil of the great Thomas Telford—that set his youngest son on the path of civil engineering.[8]

4

THE EXACT DATE when Charles and his sister were reunited with the family in Waterford is not known, but it was likely to have been when Charles was about 11 years old (1854), when he was ready for formal schooling.

John O'Connor's choice of school for his second son is a puzzle for a biographer because no record of Charles's schooling has survived and he himself makes no mention of his schooldays in his extant papers. Tauman's otherwise reliable 1978 biography states that he went to Bishop Foy's School (sometimes known as Waterford Endowed School or the Blue School), but this seems highly unlikely. The autocratic, doctrinaire Bishop Foy founded his school in 1706 to give a basic primary education to poor Protestant boys, sons of local labourers and tradesmen, and left sufficient money in his bequest to apprentice them in the city on completion of their meagre schooling. Conditions at the school, even in O'Connor's time, could scarcely have been much better than at Dickens's Dotheboys Hall. The Endowed School Commissioners, at their 1855 public inquiry into the school, were told of the extreme dirt in the corridors and dormitories, of the children's neglect, and of their bare existence on poor quality rations. Conclusions on the standard of education were no less critical:

> The general state of secular education was unsatisfactory. Only one
> boy could answer in the first book of Euclid. And in history, as well
> as in most departments of education, but little progress appeared to
> have been made, the deficiency greatest in English diction.[1]

This judgment hardly equates with what we know of O'Connor's high scholastic ability, his literary facility and his broad general knowledge.

[33]

Given that John O'Connor, JP, had a position of some importance in the community and a good, if not opulent salary of £300 per year, we can assume that he selected a school that better reflected the status of the family and provided a quality education for his sons. But perhaps the most convincing evidence that the young Charles did not attend Bishop Foy's School is the absence of his name in the detailed registers of the school for that period, which are preserved in the Representative Church Body Library in Dublin.[2]

So where did young Charles receive his formal education? With little doubt, at the same school that his elder brother George attended, the Waterford Academy.[3] Their father's salary would have enabled him to pay the modest fees for day scholars of 4 guineas a year, and the family would have lived, if not lavishly, at least comfortably. A myth in some accounts of O'Connor's life has it that the family, ruined by the Famine, lived in near penury when they moved to Waterford.[4] But John O'Connor's salary at that time was considered adequate for 'a gentleman, his lady, their three children and two maid-servants'.[5] A typical annual budget comprised household expenses amounting to £161 17s per year; rent and taxes, £30; education, £12; and wages for the servants, £31. According to the personal investigations of author and philanthropic journalist Henry Mayhew—admittedly relating to London but in the same period—food was cheap: butter and tea, 5d per pound; meat, 1s for 3 pounds. Herrings were sold at three or four for a penny and potatoes (after the Famine) at 2d per pound.[6] If anything, prices of these commodities in Ireland could have been even lower. Seen in this light, John O'Connor's salary (possibly supplemented by rent from 'Ardlonan') would have kept the family in reasonable comfort.

Waterford in the mid-nineteenth century boasted several schools, but the Waterford Corporation Free School, otherwise known as the Waterford Academy or the Latin School, was generally superior to the others, having much in common with the old-style English grammar schools. Pupils were drawn from middle-class families of the Established Church, many of whom went on to Dublin University. The school's purpose was to provide a sound classical education for the sons of the city's professional and business classes. Subjects taught included English Literature and English Grammar, Mathematics, Bookkeeping, Greek, Latin and French, Music, Drawing, and extras like Dancing and Fencing. Charles must have been a bright pupil. His grasp of mathematics and geometry—clearly essential for his future career—was exceptional. Most likely he was a studious boy and, as we know from his later correspondence and brief diary jottings, he read widely.

In later life, O'Connor made no reference to his education, commonly a time of youthful hardship and adolescent cruelty. But under Charles's headmaster, the Reverend Dr William Price, the academy enjoyed a high scholastic reputation 'equal in rank to any similar establishment in this kingdom'.[7]

John O'Connor's salary was evidently not sufficient to enable his sons to enter Dublin University (thus breaking a family tradition), but he could well pay—if it were required—the premium for Charles's pupillage to John Chaloner Smith, successor to G. W. Hemans, chief engineer of the railway company.[8]

Why Charles chose civil engineering is easy to understand. Not only was he living in an age when the great achievements of engineering would have excited the imagination of all adolescents, but in addition he would surely have met and been inspired by the engineers who worked closely with his father on the railways. Boys of those days were as fascinated by the technology of the age as boys are today—as testified to by a German visitor to Manchester Grammar School in 1844, who observed that the favourite subjects among the graffiti on the desks of the pupils were 'canals, railroads and rivers with barges, ships and locomotives on them'.[9]

John Chaloner Smith, Charles's master (and soon to become his mentor and friend), was one of those cultured Victorian polymaths who seems to have made an equal contribution to civil engineering and the arts. After gaining his Bachelor of Arts at Trinity College, Dublin, in 1849, he became articled to Hemans and was promoted to resident engineer to the Waterford & Limerick Railway Company in 1853 on the recommendation of his master. When not engaged on problems associated with railway building, he was studying engravings and mezzotints, becoming a world authority on these media, and amassing a collection of mezzotints that later found its way into the Dublin National Gallery and the British Museum. His study *British Mezzotinto Portraits*, in four volumes, published in London, 1871–82, is still considered the authoritative work on the subject. In railway engineering, Smith was equally qualified and successful, forming his own railway construction company in partnership with John Bagnell. O'Connor remained with Bagnell and Smith after he had completed his apprenticeship, and worked on several railways, including the Boris and Ballywilliam line north of Waterford in Kilkenny, the Clara and Streamstown line in Westmeath (now no longer in existence), and the Roscrea to Birdhill line north-east of Limerick in Tipperary.

Later in life, after O'Connor had departed, Smith was recognized as the 'grand old man' of Irish railway engineering, serving as secretary of the Institution of Civil Engineers of Ireland, and president in 1893 and 1894. C. Y. O'Connor was surely fortunate to be Smith's pupil, not only because of the Telford connection through Hemans but because Smith 'had a pleasant genial manner, and was particularly kind and encouraging to his assistants, and was always anxious to recognise ability wherever he found it'.[10] Australia and New Zealand are also fortunate that O'Connor did not stay working for Smith, who went on to receive more extensive and challenging railway contracts.

It was while serving with Bagnell and Smith that O'Connor would have come in contact with Isambard Kingdom Brunel. Brunel was consulting engineer to the firm, and no mere cipher on a letterhead or company report if his stated views are any guide. He was known to despise the prominent engineer 'who for a consideration sells his name and nothing more' and was careful to give more than his name to the works with which he was associated, visiting Ireland from England on more than one occasion.[11]

O'Connor's training and early work on railway building in Ireland, in solving problems of drainage, gradients and gauges, the power and limitations of different engines and rolling-stock, rock cutting, tunnelling and building bridges, were the foundation of his engineering training; they were to make him eminently suitable for his work in New Zealand, and more particularly for his work on upgrading and designing railways in Western Australia.

O'Connor completed his apprenticeship at the end of 1861 and from 1862 was employed as one of Smith's several young assistant engineers—'the old chums', as O'Connor later referred to them.[12] One of the assistants was Cecil Darley, who had been a fellow apprentice with O'Connor and spoke of him as 'his dearest friend'.[13] Darley later became Chief Engineer in Sydney and his brother, Sir Fred Darley, Chief Justice of New South Wales.

At the time when O'Connor entered Bagnell and Smith's employ, a scandal broke in Dublin touching the Yelverton family, which, though not involving O'Connor personally, must have been a sore embarrassment to him, and perhaps the one occasion when he may have rued his Yelverton name.

Charles's first cousin, the Honourable Major William Charles Yelverton, appeared in a Dublin court towards the end of February 1861 on a charge of bigamy. The scandal came to light when a Glasgow hotel filed a claim to recover overdue accommodation costs amounting to more than £259 for Yelverton's wife, the Honourable Mrs Maria Teresa Yelverton. Defence counsel for Major Yelverton denied that the claimant was his wife, Yelverton testifying that his real wife was the widow of the late Professor Forbes of Edinburgh.

Major Yelverton was clearly a 'cad and a bounder', in the time-honoured Victorian definition. As revealed in the evidence, he had met Maria Longworth (as she then was) on a cross-channel steamer from Boulogne, subsequently seduced her in Edinburgh, and proposed marriage. He was severely in debt at the time and it was clear that he had an eye on Longworth's modest fortune, which he anticipated would clear him. They went through an informal civil marriage in Scotland but Maria, a Catholic, was not satisfied with the ceremony. Yelverton, 'in order to satisfy her scruples', consented to a marriage conducted by a priest, the Reverend Mooney of Rostrevor near Newry. Yelverton was said to have paid the priest two £5 notes. Then followed a protracted honeymoon, the couple visiting the Giants Causeway, the Scottish Highlands and then the Continent, and in all these places they booked into hotels as Mrs and Mrs Yelverton. During this time, the marriage had not been made public because Yelverton had married without the consent of his father, Charles's uncle, the 3rd Viscount Avonmore. The heir feared the viscount's disapproval, not without good reason.

Yelverton's leave expired and the bridegroom returned to his regiment, leaving the bride—now pregnant—in France. From England he wrote to her advising an abortion and declared that he was a ruined man unless she consented to go to Australia, where he would 'join her in a few months'. She shrewdly refused both of these requests and chose to return to relatives in Scotland. It was there, in Leith, that she heard of her husband's second marriage to the widow Emily Forbes.

The court case continued over nine days, revealing all the sordid details. Yelverton, under cross-examination, graphically described how he had seduced Maria Longworth, at which point 'the ladies were compelled to leave the court and the details which followed were unfit for publication'. According to *The Times*'s correspondent, the case created unprecedented interest:

> The people of Dublin seem capable of thinking or talking of nothing else. The excitement is not confined to the metropolis. All the provincial papers are full of the proceedings. Some of them omit their leaders in order to make room for the report; others say they omit advertisements for the same reason.

The jury deliberated for only twenty minutes, and on its return was unanimous that Maria's Scottish marriage and her Irish marriage had both been valid. The audience burst into applause, 'which was taken up by the immense crowds outside the courts and along the Quay'.

But the drama did not end there. The Yelverton family challenged the Irish court verdict by appealing to the Scottish court of sessions, where the Irish marriage was annulled. Maria Longworth then appealed to the House of Lords. The final judgment appeared sympathetic to Longworth, but by a majority of three Law Lords to one, the Scottish court finding was upheld, seemingly on a technicality.[14] Great sympathy for Maria Teresa Longworth was expressed throughout the nation and a subscription on her behalf was raised in Manchester to pay for her long legal struggle.

William Charles Yelverton, who had served in the Crimean War with distinction and was created a knight of the fifth class by the Turkish Government, was suspended from all military duties and placed on half-pay. He died abroad at the age of 59. Maria Longworth spent her later years travelling and writing books, including an account of the history of the case and her unhappy experiences with Yelverton.

It would have been impossible for Charles Yelverton O'Connor not to have been distressed by his cousin's disgrace or to ignore the court proceedings. It was a cause célèbre which, as *The Times* quaintly reported

> created an extraordinary demand for newspapers. It being quite impossible to supply this demand in Dublin, the printing machines were kept at work almost incessantly.[15]

We may wonder what effect the Yelverton notoriety had upon the young Charles O'Connor, and whether his cousin's caddish behaviour in any way directed his own course of action when the time arrived for courtship and marriage.

The scandal surrounding the marriages of the 4th viscount was long remembered and resented in parts of Ireland, being interpreted as yet another instance of Protestant power versus Catholic impotence. The Avonmore estates suffered a decline from that period, not least because of the voluntary exile of the disgraced heir, who succeeded to the title in 1870.[16]

To what extent the Yelverton scandal influenced the young O'Connor to seek his fortune overseas is hard to say, but it must have played no small part.

5

IN THE MIDDLE of the year 1859, John O'Connor's name disappeared from the company reports of the Waterford & Limerick Railway Company: a new secretary, Thomas Ainsworth, took his place. We do not know the reason why he left or whether he received a pension. He was 53 years old and had been with the company only eight years. His son remained apprenticed to the resident engineer, John Chaloner Smith, for a further two years.

The next we hear of Charles's father is his death four years later at the age of 57 on 8 June 1863, and his address was then given as Prince William Terrace, Dublin—a stylish Georgian, tree-lined avenue likely to be described in today's advertisements as 'comprising highly desirable, superior residences'. The house was probably owned by Mrs O'Connor's brother-in-law, the 3rd Viscount Avonmore, and let to the O'Connors at a peppercorn rent.[1] The probability that John O'Connor faced impoverishment and needed help after leaving the railway company is further evidenced by the details of his will. His widow, Mrs Elizabeth O'Connor, died at the Dublin home of her sister Anna within two months of her husband, which would account for the delay in granting probate until the following February. John O'Connor's meagre estate amounted to less than £300, and, curiously, this small sum went to his youngest (unmarried) daughter, Letitia O'Connor.[2] The eldest son, George, then serving with the Royal Artillery in India, is recorded as owner of the 'Ardlonan' estate, entailed to him several years before John O'Connor's death.[3] The whereabouts of the second son, Lorenzo, in the United States with the Waterford Artillery, were unknown and he is assumed to have been a casualty in the Civil War. But it is curious that none of his father's meagre legacy was left to Charles, the son who was probably by his side at his death. Did John O'Connor, so demonstrably compassionate during the Famine, leave what little he had to his one child who had no means of support? Or was the father estranged from Charles on account of him having chosen engineering—a career supposedly below his station, as one biographical note maintains?[4]

Like his father before him, John was buried in the churchyard at Castleknock not far from the western entrance of Phoenix Park, the graves now so worn away by the weather that little can be made of the inscriptions.

With his parents dead, and his two elder brothers in the army overseas, Charles must suddenly have felt deprived of family support. But with his apprenticeship concluded, his work at Bagnell and Smith would have occupied him fully throughout those difficult months, as well as providing a modest salary. As he recalled later in life:

> I was engaged as an Assistant-Engineer in charge of the construction of several weirs on the River Bann in the North of Ireland and had exceptional opportunity for realising the conditions and forces which had to be dealt with in flood time.[5]

His qualities were evidently recognized by his employer and mentor, John Chaloner Smith, because he was given increasing responsibility. Smith wrote later in a testimonial that O'Connor showed himself to be 'a quick and accurate surveyor, an able accountant, and as a capable manager of men'.[6]

Capable, serious, responsible—these are the characteristics evident in Charles O'Connor's formal photograph taken in a Dublin studio about that time (see opposite). From the approximate date, it seems likely to have been a portrait celebrating his completion of articles. The young O'Connor is fashionably dressed, looking dapper in black frockcoat, waistcoat, high-waisted grey trousers, and cravat. His right hand holds a cane; his left is held behind his back. His strong, firm mouth, high forehead and distant gaze foreshadow the mature introspection of the Porcelli statue. Self-confidence and leadership potential are also glimpsed in the portrait, and these qualities developed and matured in the years ahead and enabled O'Connor to employ, guide and control engineering work gangs successfully in both New Zealand and Australia. O'Connor's own opinion of himself at this time, expressed some years later, is less flattering. He describes himself as 'conceited, exclusive and pedantic' and

> a mere half-educated boy, saturated in old-world formulae, and representing merely the outcome of the thoughts of others impressed upon the elastic mind of a child...[7]

C. Y. O'Connor as a young man in Dublin, c. 1862.

Courtesy Muriel Dawkins

Charles celebrated his 21st birthday on 11 January 1864, doubtless employed somewhere in the heart of rural Ireland, perhaps sheltering from winter storms under canvas, surveying a new railway line or supervising the construction of a weir or bridge. But his thoughts, when the day's work was done, surely would have turned to a consideration of his future. It seems likely that he was given no assurance that his employment with Bagnell and Smith would continue when their current contract was completed. 'I am afraid most of the work at home is done, and what remains is too much completed for it to be worth much', he wrote to his friend Darley.⁸ As a result of the heady years of the railway boom, the burgeoning profession of civil engineering found itself oversubscribed when the demand for new lines in Ireland diminished towards the end of the 1860s.

But the demand for railways did not diminish overseas until much later. Railway fever had spread to the Middle East and the British colonies. For example, Thomas Brassey, one of the most powerful railway construction magnates of the era, after constructing railways in England and on the Continent, built some 4,500 miles of line in Canada, the Crimea, Australia, the Argentine and India, all within the space of seven years from 1862 to 1869. Given that the popular subject discussed among all young men in Ireland at the time was emigration, Charles and his engineering colleagues inevitably would have been tempted by the seemingly limitless opportunities afforded their profession worldwide. Mixed with his sense of ambition and adventure, the young O'Connor would also have been touched by the prevailing Victorian sense of 'mission': the belief that Great Britain's dominant position in the world required her to share her industrial, commercial and moral advantages with those less fortunate in her newly established colonies.

Perhaps inspired by accounts of Brassey's railway enterprises in the Argentine, O'Connor's first thought was to make South America his preferred option, as evidenced by a letter he solicited from William Malcolmson, one of his father's old friends in Waterford, later to become head of a successful industrial and railway-building company. 'As your talents are more cultivated for railway purposes than anything else', Malcolmson wrote:

> I should think the River Platte would be a better market to take them to…We have steamers monthly from Liverpool to the River Platte, on which I could give you a free cabin passage and you would have to find yourself on board or contract for your feeding with the Captain. I know no one there to whom I can give you an introduction, but I will be glad to forward to you a general character reference.⁹

But by the beginning of December 1864, O'Connor had decided in favour of going to New Zealand, as we know from a letter by his mentor, John Chaloner Smith, to G. W. Hemans in London:

> Charles O'Connor, the son of the late Secretary of the Waterford and Limerick Railway is going out to New Zealand as there is but little prospect of work here…Could you kindly tell him to whom he should apply for information and if possible for an appointment. He is really a smart and good man. Yours very truly John C. Smith.[10]

The decision in favour of New Zealand was no doubt influenced by O'Connor having relatives there: while one of his two maternal aunts, Cecilia O'Keefe, married the 3rd Viscount Avonmore in Ireland, the other, Sydney Eloise O'Keefe, married Lieutenant Colonel Foster Goring, who held the position of secretary in the New Zealand Executive Council in Auckland. The O'Keefes were related to the Yelvertons not only by marriage but through the Belle Isle Yelvertons (see endpapers, inside back cover).

And so Charles Yelverton O'Connor left his native Ireland a few days before Christmas 1864, armed with a sextant—a gift from his friend and employer, John Chaloner Smith—and several letters of introduction, including one from his uncle, Viscount Avonmore, testifying that his nephew had been active and efficient in the performance of his duty: 'I know that they [his employers] placed much confidence in him, and that he gave them general satisfaction'.[11] It was not effusive, perhaps, but in the general practice of the period, a few mild words of commendation from a member of the aristocracy—a relative—were enough to ensure that Yelverton's nephew would never want for support in the colonies.

O'Connor's choice was 'a first-rate ship, quite new',[12] the 1,183 ton *Pegasus* in the command of Captain Cornwall, which left the Downs, at the mouth of the Thames Estuary, on 24 December. He was not to see Ireland again for thirty-three years—and then only briefly.

Had he stayed on in Ireland he most likely would have worked with John Chaloner Smith, who later secured the contract for building one of Ireland's most spectacular railways, connecting Wexford with Connolly Station, Dublin, via the Liffey Viaduct. The tunnelling through headlands at Dalkey and Bray, and the masonry bridges across the Dodder and Shanganagh rivers, make up a route considered by many to be among the finest in the country.

PART II
NEW ZEALAND, 1865–1891

6

—

Pegasus, UNLIKE THE mythical winged horse that inspired its name, was a comparatively slow, three-masted ship built in 1862 at Miramachi, on the east coast of New Brunswick. It was making the first of two voyages to New Zealand, carrying fare-paying settlers. Charles O'Connor shared a cabin with six other voyagers, including a parson, a doctor and assorted babies. 'We got on first rate', he wrote, '—when the babies were asleep', hinting that their 'singing' contributed to the discomforts of the voyage. He adds that one particular cabin companion was an 'awfully jolly fellow and we became great friends'.[1]

Pegasus headed south across the Bay of Biscay and down the west coast of Africa, sighting St Antonio in the Cape Verde Islands eighteen days out, on 11 January 1865. It was O'Connor's 22nd birthday.

O'Connor was evidently a personable, friendly and eager young man himself, intensely interested in his new experiences and resourceful enough to benefit from them. With his sextant, he spent much of his time studying navigation: 'The skipper was a very good fellow…he let me work by his chronometers, and lent me charts and all sorts of things'. And he reports that they had a very lucky voyage, with no rough weather, 'so I never suffered from sea illness'.[2]

The ship's course would have taken them around the Cape of Good Hope, across the south Indian Ocean, across the Australian Bight, and then northward across the Tasman, rounding the tip of North Island to proceed south again to Auckland. O'Connor wrote that after sighting the Cape Verde Islands, 'we never saw land again until Three Kings Islands on 24th March'.[3] The voyage on the *Pegasus* had lasted three months.

When O'Connor and his fellow immigrants arrived in Auckland at the end of March 1865, New Zealand had been a British colony for only twenty-five years, and was then governed by a central parliament (soon to be moved from Auckland to Wellington), and nine distinct provincial councils that retained power over Crown lands, education, police, public works and immigration. Of the provinces, there were four in the north: Auckland, Taranaki, Hawkes Bay and Wellington; and five in the south: Nelson, Marlborough, Canterbury, Otago and Southland.[4] Under this system, New Zealand has been described as more a federation of small settlements than a unified colony.[5] A characteristic of the majority of provinces was the religious foundations of their settlements. Otago, for example, was largely colonized by Scottish Presbyterians; Canterbury, by English Protestants; Taranaki, by Methodists. The church groups in England had been active in recruiting New Zealand migrants, much as the America-bound Puritans had done in 1620. By 1861, the European population numbered a little over 99,000, with 42 per cent in North Island and 58 per cent in the south. These figures would more than double in the next ten years due largely to the gold discoveries in Otago in the south, and on the West Coast.

The earliest settlers had been traders and fearless missionaries who attempted, with some little success, to pacify the warlike Maori tribes. But even the Treaty of Waitangi of 1840 did not entirely end the conflict between the settlers and the indigenous population. In central North Island, the tribes, intelligent and well-armed, kept up partisan warfare so that colonial forces had to be maintained in the area throughout the 1860s. Not until 1871, when many of them were worn out and depleted in numbers and several of their leaders had enlisted with the colonial army, did the Maori fighters finally lay down their weapons and fighting cease.

O'Connor resided with his aunt's family in Auckland for only one week before he found work with a government survey team working in remote Ngakinapouri, approximately 68 miles south of Auckland on the Waipa River. To reach the camp, he travelled by coach as far as the Waikato River, and then by government steamer upstream until meeting the confluence with Waikato's tributary, the Waipa, which occurs just below the site of the modern city of Hamilton. William Pember Reeves describes the Waikato as 'the longest and on the whole most tranquil and useful of rivers traversing a curious region of hot geysers, mountain lakes, streams, and patches of forest'.[6]

O'Connor's survey camp comprised four tents and a cookhouse. 'Besides the boss and myself, there are nine men and a cook and we get on first rate.'[7] The only hint of danger from Maori attack that he allows himself to give is a reference to soldiers in the vicinity and his explanation that the survey team was supplied with the same free rations that were allocated to the troops. He then writes:

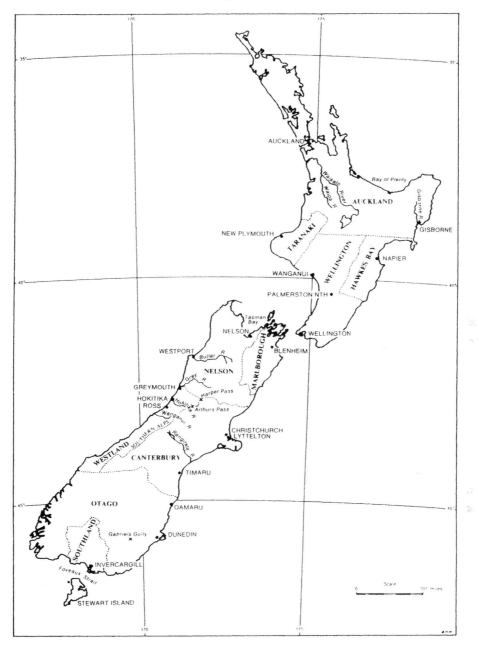

New Zealand, 1865–91.

Map by Geoff Ward (1978)

the only hope of holding the country lately confiscated from the natives is packing it all over with militia and military settlers or keeping a standing army of 10,000 men.[8]

The reference here is to Governor Grey's proclamation, the previous year, providing for the confiscation of 3¼ million acres in the Waikato, Taranaki and Bay of Plenty areas. As a consequence, the Maori wars were far from over in the months when O'Connor was working in the bush in the Waikato region, although the main areas of guerilla warfare had moved further south to the Taranaki and Wanganui districts, and further east to the Bay of Plenty. If the government surveyors were in any danger, it was more likely to come from the Hauhau cult—in name and barbaric cruelty precursors of the African Mau-Mau. Hauhau, which was active in the Wanganui district in 1865, was a savage expression of a revolt against Christianity and Western civilization and drove its adherents to acts of pitiless and ferocious cruelty. The members of the cult worked themselves up into a frenzy by means of dances and incantations, prior to launching their attacks directed not only against Europeans but also against neighbouring tribes friendly to the settlers. 'Until the present crisis is over I would not recommend any fellow coming out', wrote O'Connor.[9]

O'Connor enjoyed his rough life in the bush, 'the most free and easy style imaginable'. 'I won't deny it's hard work', he wrote, 'perhaps up to your knees in water all day—of course not every day—but it is a jolly independent life'.[10] Surveying was contracted out by the government, and in this case the contract, covering between 20 and 30 miles, was held by a young man from County Cork named Long—'a very good fellow', only one year older than O'Connor himself. Long paid O'Connor £1 per day as his assistant.

> I did not like to take a contract by myself at first both because I had not sufficient capital to carry it on properly and also because I wanted to know more about the ways of the place—but when this job is finished I expect to be able to get one on my own or go into partnership with Long.[11]

But freelance contract work and partnerships were not the path that O'Connor ultimately followed when the work was completed in the Waipa area. He stayed in company with Long, however, and together they moved to the South Island, where they joined the engineering and surveying staff of the government of the Province of Canterbury.

The naming of the province by the first settlers in 1850 was a statement of their aspirations and intentions. When they arrived in December 1850 on four ships, the 'Canterbury Pilgrims', as they chose to call themselves, dreamed of creating a model Protestant society linked with the symbolic centre of their church, Canterbury Cathedral, in southern England. The capital they named Christchurch, and the nearby port, Lyttelton, after George William, Lord Lyttelton, Under Secretary of State for the Colonies and chairman of the Church of England corporation that was responsible for establishing the settlement.

At the time of O'Connor's arrival in September 1865, Canterbury extended across the country from the east to the west coasts. There the young surveyor would have encountered a land markedly different in climatic and physical characteristics from those he had experienced on the Waipa River. From Christchurch, the administrative capital on the eastern seaboard, Canterbury's grassy, pastoral plains extend westward for roughly 43 miles until they meet what appears to be an insuperable barrier: range upon range of precipitous mountains and gorges stretching north to south of the island like a giant backbone. These are the rugged Southern Alps, snow-capped for most of the year. On the east side of the range, the foothills are covered in dry tussock grasses that give way on the higher slopes to dark grey, bald rock curiously marked by lighter patches running down the sides like treacle down the side of a pudding; these are smoothed courses created by retreating snows. Although the eastern slopes have a light rainfall, the valleys and the Canterbury Plain are well watered by treacherous, fast-flowing rivers rising in the mountains. These watercourses are generally wide valleys—seemingly harmless ribbons of water cutting their way between islands of rock and shingle. But after heavy rains in the mountains and melting snows, the ribbons can suddenly become perilous torrents that can maroon the unwary traveller on the central islands:

> The rivers were seldom fordable on foot, and then only when they were very low, as in winter. To ford safely great care had to be taken to enter the water just above the ripple in the fall, following the shallow water and avoiding the deeper channels. Frequently a very shallow-looking fall might have a deep gutter on the far side in which a horse would have to swim. A ford such as this is very treacherous, and has been the cause of many deaths.[12]

As dangerous as these rivers were before the construction of adequate bridging, they were—and still are—the source of life irrigating the dry but fertile Canterbury Plain famous for its rich agricultural tradition.

However testing O'Connor's bush life might have been in the north, it was merely a prelude to the privations and dangers that he would face in his new employment—dangers not from Maori warriors, whose presence in the south was minimal and generally friendly, but from the harsh, largely unexplored terrain itself. 'More than one of the Government officers sent there [the Alps] were either swept away by some torrent or came back half crippled by hunger and rheumatism.'[13]

O'Connor's fast developing career in New Zealand owes much to the discovery of gold there, and the subsequent gold rushes on the West Coast, which roughly coincided with his arrival. It is interesting to note that just as gold was the catalyst for his advancement in New Zealand, so too were the major gold discoveries in Western Australia the foundation of his work there in the 1890s. In New Zealand, the first major finds—alluvial gold in the river beds—were discovered by the prospector Gabriel Read, in the neighbouring Province of Otago, in 1861. As a result, Presbyterian Dunedin became the fastest developing city in the south, and sober diggers from there, from Australia and from elsewhere converged in their thousands on the new goldfields. Three years later, gold was discovered in the Canterbury Province, on the western slopes of the Alps and in the alluvial mud washed down from the mountains to the coast; early prospectors even found gold among the shingle on the beaches. A similar rush followed the news, and parties of diggers transferred their allegiance from Otago and poured into the neighbouring province. They faced such harsh conditions that many of their number starved, or were lost in the bush, or drowned in the fast-flowing torrents. A significant number of prospectors came by sea from Australia and landed on rocky, dangerous shores near the mouths of the Hokitika and Grey rivers. The town of Hokitika, from its humble beginnings as a collection of tents pitched on the narrow coastal strip in 1864, grew within a year to be one of the largest settlements on the South Island.

> Captains of vessels sailing along the coast saw a continuous line of fires each denoting a party of miners. All along the beaches to Greymouth and up the Grey river, and from there to Teremakau, the busy hive of workers could be seen.[14]

By 1866, gold exports from the area amounted to in excess of £2,000,000. The population of the province, 16,000 in 1861, more than trebled in the next ten years.

With the growing importance of gold, and the resultant general prosperity and increase in population, the provincial government was under mounting pressure to improve communications with the West Coast, hitherto virtually cut off from the east by the mountain range. The first writer to describe the West Coast was Captain Cook, in 1778: 'An uninhabitable shore of rocks covered with snow, and as far as the eye could reach, the prospect was wild, craggy and desolate'.[15] A hundred years later, a similar view was expressed in the provincial council:

> The dangerous character of the harbours of the west coast and the great destruction of shipping which has taken place…make it clear that the trade of the Goldfields cannot be carried on with reasonable security until a road fitted for heavy traffic has been completed connecting the eastern with the western side of the intervening range.[16]

Also recognized as similarly important for trade and immigration was the provision of safe harbours on the storm-lashed western seaboard.

A rough track that already existed across the mountains via the Otira Gorge had been originally created by the Maoris, to enable them to bring their precious greenstone to the more populated areas in the east. But it was unsuitable for coaches and drays carrying vital supplies, equipment and mail. In the early period, stores were mostly brought in by ship. Clearly, the provincial government could not delay indefinitely the provision of a safer and more direct access to the West Coast. To this end, Edward Dobson, Provincial Engineer for Public Works, was charged with the task of upgrading the road, and to accomplish this he expanded his department. Among his new employees was the young C. Y. O'Connor, who was assigned to that section of the road named after one of Dobson's two sons, the infamous and still celebrated Arthurs Pass, 7,457 feet above sea level.

O'Connor's appointment as assistant engineer commenced on 6 September 1865.

7

ALTHOUGH ARTHURS PASS, which leads on to the Otira Gorge, with Mt Rolleston (7,454 feet) on one side and Mt Franklin (6,853 feet) on the other, is not the only route across the mountains from east to west today, it was the first one established and is the most celebrated and impressive. Not the highest point in the Southern Alps—that honour belongs to Mt Cook at 12,321 feet, some 80 miles further south—the area is designated a National Park and is visited for its scenic grandeur. Its flora and fauna are avidly photographed by admiring tourists, who travel either by a modern, safe but precipitous road or by the trans-alpine express from Christchurch to Greymouth on the West Coast. There they can visit the local history museum and the reconstructed goldmining shantytown. Few cars and buses would pause at the narrow Taipo River Bridge west of the gorge, where a modest plaque, erected by the New Zealand Institution of Engineers, commemorates the engineering work of C. Y. O'Connor—in particular, the foundations of the bridge itself, which were designed by him to carry the railway. It was O'Connor's pioneering survey back in 1865–67 that is largely responsible for the tourists' safe and comfortable crossing of the mountains today.

The Otira Gorge was anything but comfortable and safe when O'Connor first journeyed there in September 1865 in command of a small party of assistants, horses, tents, supplies and sundry equipment. It was a grim, dangerous assignment in uncharted country ruptured by swift-flowing icy torrents, precipitous ravines and, on the lower slopes, dense rainforests made impenetrable by bracken, fern and creeper. The higher slopes were snow-clad in the winter months, and at other times bleak, treeless and windswept. On the westerly side of the ranges, in many places bordering close to the coast itself, the rainfall totals 100 inches a year; it is a wild and mysterious world of lush forest, with giant rimu and rata trees towering over luxuriant lichens, mosses

Otira Gorge, the precipitous east–west road across the New Zealand Southern Alps. An artist's impression c. 1866, shortly after construction, following C. Y. O'Connor's surveys in 1865–66. *Courtesy West Coast Historical Museum, No. 1712*

and ferns. The ferns grow some 23–33 feet tall in their struggle for light. In such country, the movement of men and horses was regularly impeded not only by the growth but also by heavy rain, and accomplished only with the greatest difficulty through the thick mud and debris washed down the mountain sides. Countless tales are recorded of the privations and dangers suffered by early explorers and surveyors and of their difficulties in traversing the country. In one recounted by Edward Dobson, who employed O'Connor in similar terrain, a digger camping near the beach shot a pigeon in a tree a short distance away while standing at his tent door. The man pushed through the scrub to retrieve the bird, but it was so thick that he had to crawl on his hands and knees under masses of vines. It took him until dinner time to reach the tree where the pigeon had been shot and until dark to get back to his tent—a distance of about 22 yards.[1]

Dobson's colourful account of his survey work on the West Coast is a valuable guide to the kind of life that O'Connor would have faced there. Edward Dobson, a Canterbury Pilgrim, explored the same area before being appointed Provincial Engineer and, once appointed, employed O'Connor, one of five surveyor-engineers including his own son, George. Doubtless he would have instructed and influenced his new employee, giving O'Connor the benefit of his wide experience. On his treks, Dobson had employed Maoris to assist him and compared them favourably to 'white men', who were, he stated, 'quickly disheartened'—as well they might have been. Seven men, he states, were drowned on the West Coast surveys in as many months.[2]

However, there were compensations, especially in the summer months. The slopes of the Alps are an area of great natural beauty, as so memorably described by New Zealand's pre-eminent historian, William Pember Reeves:

> All around is a multitudinous, incessant struggle for life; but it goes on in silence, and the impression left is not regret, but a memory of beauty. The columnar dignity of the great trees contrasts with the press and struggle of the undergrowth, with the airy lace-work of fern fronds, and the shafted grace of the stiffer palm-trees. From the moss and wandering lycopodium underfoot, to the victorious climber flowering eighty feet overhead, all is life, varied endlessly and put forth without stint.[3]

Sadly, the diary that O'Connor most assuredly would have kept in 1865 while on this assignment has not survived, but we get an idea of what his life must have been like from descriptions in his diary for the year 1869, written when engaged on similar work based out of Hokitika.[4] Although he confines himself in that instance to reporting simple facts, eschewing drama and descriptive prose, the simple day-to-day, understated repetition gives an impression of O'Connor as a hard-working, practical and patient young man, respected by his team but in command. He rises early every day, as soon as it is light, either he lights a fire, or one of his men—the cook—does, and he has breakfast, usually porridge. Often it is too wet for a fire and he makes do with cold soup from a tin. He enjoys his first smoke of the day. His concern is always the weather, which determines how much triangulation he can accomplish. Often he notes that it is too cloudy—'no chance of shots'—or that he 'could not see hills for a long time and then only got one dim sight'. The rain seldom restricts his ability to move about, and although it is clear that his trekking must have been severely uncomfortable at times, there is no hint of frustration or ill-temper in his writings. Occasionally, he is confined to his tent all day but he uses this time to write his diary, transfer his readings to his 'level book', work on his accounts, and write his reports. In the 1869 diary, we learn that he carries a supply of good books with him and reads in his tent on Sundays, generally a rest day, or at times when it is impossible to move about outside.

There were dangers of another kind in the bush, which a diary of this period might have mentioned. Although the historian-politician William Pember Reeves, in his colourful account of life on the West Coast goldfields at that time, states that 'there was little crime or even violence',[5] Edward Dobson tells a different story—he having very personal reasons for doing so.[6] His son George, a surveyor-engineer and close colleague of O'Connor's, who must have been working in an area close to O'Connor, went missing on 24 May 1866.

> The father of the lost man is in a fearful state about his poor George;
> he came from Christchurch by last coach, and left for the Grey
> yesterday to promote the search.[7]

The search party, which included Maori trackers (was O'Connor involved?), eventually found George Dobson shot and left to die in the bush near the Grey River. A man named Sullivan, a member of the so-called 'Kelly gang', turned King's evidence and confessed to this and several other murders of miners thought to be carrying gold from the diggings. George, not carrying gold, had been killed, like other unfortunates, because he had recognized the men as outlaws. This was not an isolated case. Edward Dobson states that when on road work on the West Coast, he always carried firearms 'in consequence of the hold-ups and murders that had been going on'. He describes how he always carried 'a revolver in a holster and also a double-barrelled pistol (a derringer) in my coat pocket'.[8]

As O'Connor and his team moved westward across the mountains, determining the path of the new road and marking the route, road-gangs were moving up behind, cutting trees, levelling and grading, so that sections of the road were passable before completion of the whole. At an early stage, O'Connor formed a close friendship with the road overseer, Emanuel Rawlings, which lasted several years until Rawlings's death. We know little about Rawlings except the remarkable fact that after his tragic death at the Taipo River in December 1887, well after O'Connor had left the area, it was learned that he had left all he possessed—a sum totalling £3,000—to O'Connor in his will. A small fortune in those days, it was described in a newspaper report as 'a windfall for Mr C. Y. O'Connor' who, when the will was published a year later, was Under Secretary for Public Works in Wellington.[9] What O'Connor had done to deserve the legacy we can only speculate on: whether he had helped Rawlings, perhaps even saved his life on an earlier occasion, or lent him money, or whether it was merely a mark of Rawlings's regard for O'Connor, remains one of the unsolved, intriguing questions of O'Connor's life.

Rawlings's death provides yet another example of the hazardous nature of the work on the road in those times. It occurred on 26 December 1887, when he was supervising the felling of timber above the road, 39 miles east of Hokitika. His boot caught in the scrub as a tree fell; he was trapped, and both his legs were crushed. The injured man was in great pain as his men placed him on a spring cart and took him to Kumara, the nearest town, a torturous, jolting, three-hour ride away. Rawlings died on the journey, according to the coroner, 'from shock to the system'.[10]

In October 1866, the chief of the Engineering Department, Edward Dobson, reported to the provincial parliament that the road had been opened to traffic in March of that year, and 'since then has been metalled, widened and improved'.[11] Although O'Connor's work as a surveyor, with that of his colleagues, was evidently successful—as witness the route he selected being the one in existence today—the same might not be said for the quality of the road surface, which Dobson praised in his report to parliament. The Australian Cobb and Co. coaches were frequently bogged, and there are stories of passengers who had to leave the coach to help push the vehicle through culverts and over sections of the road that had washed away. A graphic description of conditions on the road has been left to us by Lord Lyttelton, after whom the harbour at Christchurch was named. He travelled to the West Coast with his son in 1867, a year after the road had been opened to coach traffic:

The roads, or no roads over which we passed during a great part of the journey, both going and coming, are indescribable. Through the forest it is a fair shingly road, and so it is in the high passes; but below it constantly goes through fords in the rivers and creeks, of which we must have passed some forty or fifty, often above the pole of the coach in depth and both in these and in the dry river beds which also abounded, and tractless bits of rough ground the incessant jolting was marvellous, giving quite a new impression of that marvellous process…the deep fords often seemed as if they would carry us away down stream, but the strength and patience of the horses never failed.[12]

Even today, travelling the modern road surface, one can sympathize with Lord Lyttelton. Cascades of water gush off the mountains, splashing over the road in Otira Gorge like waterfalls. Small, fast-flowing rivulets gouge out channels at the roadside. Work is constantly in progress repairing washaways and clearing rocks and other debris deposited by the force of water blocking the precipitous road surface. Proceeding with deliberate caution in a modern-day vehicle in these conditions, one can only marvel at the courage and skill of the Cobb and Co. teamsters, and at the patience of the seventeen passengers huddled together inside each rocking coach.

O'Connor's work as an engineer-surveyor at that time was evidently appreciated by his chief, who wrote later:

I have pleasure in bearing testimony to the trustworthy character of these surveys—and the intelligence and industry displayed by Mr O'Connor during the time he was engaged on my staff.[13]

O'Connor remained on Dobson's staff when the road was finished, and was engaged on further survey work in preparation for a railway to link Christchurch with the West Coast. His service record shows that in March 1867, he worked on the survey for the Hokitika–Christchurch Railway and subsequently on the railway that runs along the narrow coastal strip from Hokitika to Greymouth.[14] This was the kind of work that he had been doing in Ireland with Bagnell and Smith, but now, in the harsher, more isolated conditions of the Southern Alps, he was building up unassailable experience in surveying and planning road and railway routes, designing bridges, culverts and tunnels, keeping accounts and managing staff. In August 1867, he was appointed assistant engineer, Land and Works, for the whole of the Westland region, and when Westland was formally designated a county separate from Canterbury, O'Connor was appointed District Engineer.[15]

His diary record for the year 1869 shows him dividing his time between supervising his survey team both in the mountains overlooking the Tasman Sea and along the coastal strip, and running an office in Hokitika. He is then aged 26 and very much in charge of the preparations for construction of the Hokitika–Greymouth Railway. He makes journeys 'up the hill' usually early in the mornings, and sometimes stays several days in the bush with his men, regularly returning to his office in Hokitika to process his readings, make his calculations, keep his accounts and write his reports. On Sundays, if in Hokitika, he attends church and frequently lunches or dines with the district surveyor, Malcolm Fraser. Fraser was O'Connor's chief and a friend who provided much-needed family hospitality. O'Connor was not the only young surveyor to benefit from Fraser's friendship; Gerhard Mueller, another young surveyor, wrote to his wife in Invercargill:

> Fraser is a married man, an exceedingly nice fellow, and you may rest assured that we have tea and not beer or whiskey for tea…his wife is very nice and I spent a jolly evening talking, reading, and playing with the eldest child, a pretty girl of five years.[16]

Throughout O'Connor's diary for 1869, he notes frequent visits to Fraser's house both on business and for hospitality. At a later date, Malcolm Fraser would play a pivotal part in determining the course of O'Connor's career.

Occasionally, O'Connor frequents the old Empire Tea-rooms in Hokitika for his breakfast. On 6 February, he has a swim and then goes rowing in a small boat with 'Charlie' (surname not noted). But on 16 February he is in the bush again and 'had a very dangerous climb and very nearly came to grief'.

Most of the diary is merely a record of sights taken—or the inability to do so because of thick weather. The writing is hurried and often illegible and the daily events more often than not have been written up in retrospect. Occasionally, we get a glimpse of the more private man: on Sunday, 23 May, he is confined to his tent by heavy rain and 'reads Macaulay's essays all day', and notes that he is still reading them several days afterwards. What comes through from a detailed and sometimes difficult deciphering of the cryptic record of work from hour to hour is O'Connor's diligence and his extraordinary energy. Apart from the Sundays spent reading in his tent, his visits to church in Hokitika, and his occasional social meetings and meals with friends and colleagues, every hour

is committed to the business in hand. Seldom does the weekend afford any rest.
For example, on Saturday, 14 August:

> Up at about 7.0. Started up Crull. Got to Murray's camp about 11.0.
> Had breakfast. Went on to end of his lines about 3.40 and brought
> men back. Moved Murray's camp back to old place at commence-
> ment of his contour line and then went down Crull with Murray to
> try and find men and bring them up. After dinner went over accounts
> with Murray and made out charges and time sheet.

There is no rest the following day:

> Very wet day. Up at 7.30. Start at 9.0 down river and met men at
> about one mile below confluence of Smythe's Crull—found I could
> not get to Smythe's camp without a spurt so after a weary day's march
> and search found the tent. Got in at 6.0 and after dinner went over
> the accounts with Smythe. Bed at 11.30.[17]

It is worth noting at this point in the O'Connor story an intriguing
historical coincidence that has about it what Jung would have termed
'synchronicity'. While the young O'Connor was gaining experience as a field
surveyor, wrestling with his calculus, his geodetic triangulation and the
complicated plotting and mapping in preparation for road building in the New
Zealand bush, another young and fledgling surveyor was doing the same kind
of work at the same stage of his development, and in similarly inhospitable,
albeit flatter terrain, 4,000 miles to the west in another remote colony.
Unknown to each other then, the two surveyors—John Forrest and C. Y.
O'Connor—would later share friendship, professional respect and understanding, and would have a major impact on each other's careers.

By Christmas 1869, O'Connor, close to 26 years old, is ready for two days'
complete break from work. On 25 December, he rises at 8.00 a.m. and
breakfasts at 9.00. He attends church at 11.00 and then visits the hospital to see
a friend (could this have been Emanuel Rawlings?). He spends the rest of the
day with friends. On Boxing Day, he rises at 10.00 and spends the morning
reading *The Epicurean*, Tom Moore's satirical novel, which he finishes about
4.30. Then he goes 'up to tea to Mr Fraser's'. Later, back home in bed, he admits
to going to sleep over Macaulay.[18]

There is much evidence that O'Connor, like many of his fellow Victorian
engineers, was well read on subjects outside his own profession. As one example,

in his speech delivered to the Hokitika Town Council in March 1880 (see chapter 9), he alludes to Socrates, Plato, Galileo, Columbus, Rousseau and Napoleon among others, which shows that he had, at the very least, a passing knowledge of the history and biographies of these figures.[19] His erudition was also well known to his colleagues and friends, as best illustrated by a tribute to O'Connor written by William Rolleston, prominent Canterbury politician and one-time Federal Minister for Lands. At the time of O'Connor's death, Rolleston wrote that

> he combined gentleness and amiability with force and vigour of character and intellectual activity to an extraordinary degree...he was a deep and advanced thinker on other than professional subjects, but he never obtruded his opinion or posed as a pamphleteer.[20]

8

O'CONNOR'S SUCCESS IN New Zealand in the early 1870s owed much to the fiscal policies of the spectacularly successful and visionary parliamentarian Sir Julius Vogel. As Treasurer, and later Premier, in the central government in Wellington, Vogel pushed through a ten-year development program of immigration and public works that was largely financed by huge loans raised in London—loans then considered injudicious by Vogel's critics, who were numerous. In yet another remarkable parallel, twenty years later a similarly forceful, visionary Premier in Western Australia, John Forrest, would raise £1,000,000 on the London money market—the largest loan in Australia in those times—to finance public works and development that in turn were responsible for the final chapters in O'Connor's engineering career.

Julius Vogel was born in England in 1835, educated at University College, London, and migrated first to Australia and then to Dunedin, where he started a newspaper, the *Otago Daily Times*. He became a member of the Otago Provincial Council in 1862 and a year later was elected to the House of Representatives, becoming Treasurer in 1870. Like Forrest in Western Australia, Vogel was imaginative and far-seeing, and believed that the only way of boosting a sagging economy, and avoiding depression, was not frugality and caution on the part of the government—a conservative view—but by adopting a bold plan of increased expenditure from loans, which would, in turn, provoke immigration and general optimism. By sheer force of his personality and, it must be added, with the cooperation of a compliant Premier, Sir William Fox, the Treasurer was able to push his plans through parliament. The loans were raised or promised and thus the country entered upon a rare period of economic progress from which, coincidentally, O'Connor's career closely benefited.

While these weighty matters were being conducted on the national stage, a somewhat less significant and totally unexpected occurrence on the West Coast also favoured O'Connor's advancement. The Premier, William Fox, was visiting Greymouth in February 1872 at the same time as a major flood hit the district. On the Premier's authority, an immediate advance of £4,000 was made to the Greymouth Corporation for construction of a new harbour wall. Also, increased funds were allocated to the trunk road between Greymouth and Hokitika. And yet a further allocation was made for the commencement of the Brunner Railway between Greymouth and the coalfields, at an estimated cost of £70,000. All these works and more came under the direction of O'Connor, so recently promoted to the post of District Engineer for Westland Province.[1] The inflow of capital from the central government weakened the role of the county council and, indeed, was to hasten the end of the provincial council system—a move strongly advocated by Julius Vogel. Because the capital flowed from the central government, O'Connor became responsible to the Colonial Engineer-in-Chief and sent his regular reports on the progress of the various schemes to Wellington and not to the council. O'Connor, then aged 28, after six years in the field, had full control over local employment, letting contracts and supervising the work in hand.[2] From this time forward, he assumed an executive role that would grow in importance in the coming years in both New Zealand and Australia.

O'Connor's reputation for efficient and reliable work in Westland spread beyond the boundaries of the county. Towards the end of 1872, he was offered the more responsible position of District Engineer for the Canterbury Province under John Carruthers, Engineer-in-Chief.

When the time came for him to leave the West Coast and return to Christchurch, the citizens of Greymouth presented him with an illuminated address testifying to their high regard for him and appreciation of his work among them. Dated 16 December 1872, the formal script records O'Connor's 'unselfish and voluntary exertions' and goes on to state the citizens' desire to

> record our testimony to the general disinterested energy and ability
> with which you have performed the duties of the responsible office
> held by you while residing in this district, and our regret that one
> whom we have learnt to esteem, both in his private and public
> capacities, should be about to leave us for another part of the colony.[3]

The use of the word 'disinterested' is significant, and is probably a veiled reference to O'Connor's skill in handling jealousies and disappointments among

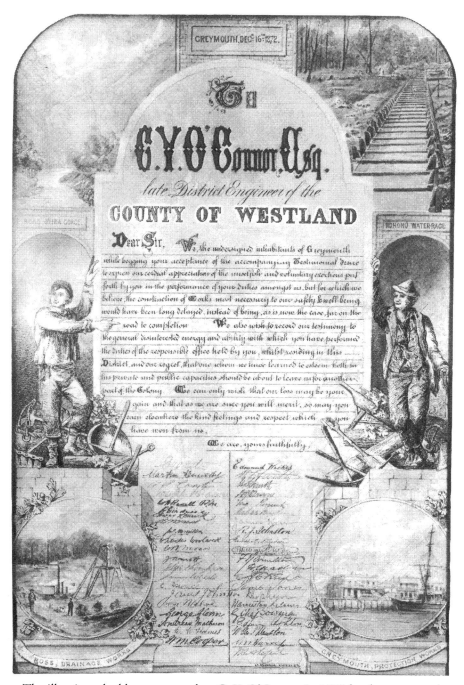

The illuminated address presented to C. Y. O'Connor in 1872 by the mayor and councillors of Greymouth, New Zealand, in recognition of his engineering works on the West Coast. *Original in possession of Judge V. J. O'Connor*

local contractors. Local patronage and jobbery were said to have been reduced under O'Connor's reign.[4] The certificate is edged with illustrations individually recording his major works: the Otira Gorge in the top left-hand corner and the Hohonu Water Race in the top right-hand corner; below these, figures representing mining; a sketch of a derrick and pump, representing the Ross Draining Works, on the bottom left; and Greymouth Harbour on the bottom right. The testimonial includes forty signatures, presumably leading citizens and members of the council. The whole design is pleasing and the artistry of a remarkable standard given the limited resources on the West Coast in those times for anything but the most practical work in hand.

Although O'Connor's new position as District Engineer would have required him to travel over the Canterbury Province, he would, from January 1873, be based in Christchurch, a growing city and important administrative centre with a population (1876) of 12,815 and an estimated 10,000 more in the surrounding suburbs.[5] From its earliest beginnings as a frontier settlement on swampy land in the 1850s, Christchurch, twenty years later, had taken on the appearance of a worthy provincial capital boasting a growing number of classical and Neo-Gothic public buildings, banks and churches. True, the Anglican cathedral in the central square, based on designs by Gilbert Scott, was only foundation high, but the wide streets were lit with gas lamps, bridges over the Avon River were in place, and the handsome Christchurch Club 'for gentlemen'—O'Connor became a member—dominated Latimer Square. The railway, linking the city with the Banks Peninsula through a new tunnel, 2,860 yards long, under an extinct volcano, eased communication with the port of Lyttelton.

For O'Connor, whose primary experience of New Zealand to date had been wild and isolated country, living in the city must have been a welcome contrast, providing more opportunity for leisure than his work on the West Coast had done. He was able to take part in Christchurch's thriving social life. More important, perhaps, he was suddenly close to key people in government and heads of departments. He would be noticed by the people who mattered. Some, like William Rolleston, superintendent of the provincial council and later a central government minister, and John Carruthers, Engineer-in-Chief, admired him and became personal friends; the latter maintained close contact and a professional relationship while O'Connor was in Western Australia. Another colleague with whom he would have worked closely was William Ness, a government architect.

Ness, Scottish-born, had come to New Zealand with his wife and three daughters in 1858 and settled in Gloucester Street, Christchurch. The Ness girls were educated privately and became active members of the local Anglican church of Avondale, singing in the choir.[6] As they grew older, they took part in the social life of Christchurch, and Susan Letitia, the eldest daughter, attended the ball given in honour of Prince Alfred, Duke of Edinburgh, second son of Queen Victoria. The prince was on a world tour in command of his own ship, the *Galatea*, visiting Australia, New Zealand and other colonies throughout 1867 and 1868.[7]

We know little of O'Connor's courtship of Susan Letitia. Probably they attended the dances that were then fashionable social events arranged in the Anglican church hall. O'Connor was tall and athletic-looking, and wore a small moustache and a finely trimmed beard, which suggested a serious, kindly nature. He had a good position in government service, with prospects of further advancement. He was well connected through the Gorings and the Yelvertons. All of these attributes would have made him an admired man and, for the daughters of professional middle-class settlers, a desirable suitor.

Susan Letitia's nature is less clear. In the only photograph we have of her at that time, she is standing behind the chair in which O'Connor sits cross-legged (see next page). She has a small, trim figure, a fair complexion, and a serious, somewhat dreamy expression. Her wavy hair is swept back from her face in ringlets, and her full, richly decorated silk dress, with a bustle, is in the height of fashion. There is some evidence to suggest that she was fun-loving, even flighty, and later that she had little interest in household management, which was conducted more efficiently by her eldest daughter.[8] What is firmly established is that Susan Ness and Charles O'Connor were lovers before their marriage. This is confirmed by the date on the birth certificate of Aileen, their eldest daughter: 4 March 1874, the day before the marriage of her parents on 5 March in the same year.[9] The bride's mother and father were the only witnesses at Holy Trinity Church, Avonside. If the dates on the certificates are correct, and the photograph described is a wedding photograph, Susan's somewhat frail and distracted appearance is readily explained. There seems little likelihood of a mistaken date on the birth certificate. If it had been deliberately falsified, it would surely be more likely to have been postdated after the marriage. Our curiosity over how Susan managed to give birth on one day and attend her own wedding on the following day cannot be satisfied so many years after the event. One explanation might be that the baby was born prematurely—or late—and the urgent wedding had already been arranged. We shall never know.

However, the question remains: was the O'Connors' marriage precipitated—even demanded—by the pregnancy? This was a common occurrence in those days of public propriety. Aileen's younger sister, the artist Kathleen (Kate)

The newlyweds, C. Y. and Susan Letitia O'Connor, 1874, in Christchurch.
Possibly a wedding photograph.

Courtesy Battye Library 67786P

O'Connor, seemed to believe so, if her somewhat acerbic letter written in eccentric old age is taken as evidence. Addressed to her sister Biddy, and undated, she scrawls that 'father did not mean to marry until he saw his little daughter and fell in love with her and adored her ever after'. She then suggests that William Ness was hostile towards his new son-in-law, and tried to 'upset his work'.[10] These comments must be judged in terms of what we know of Kathleen O'Connor's character in her eighties in Western Australia. Although she was idiosyncratic, given to caustic and often hurtful judgments, there remained at her heart's core the honesty of the artist, which requires us to consider her letter seriously. As a friend who knew her intimately explained, she could

> see truth with an intuitive, perceptive mind, but in expressing it, could seem arrogant. She had a disdain for riches and social prestige. 'What do conventions matter?' she used to say.[11]

Unlike the marriages of today's public figures and government servants, which are so often open to the public gaze, marriages of those times were intensely private and generally protected from gossip. It is useless, therefore, to speculate on the success or otherwise of the O'Connors' union. As one of Australia's leading poets has so wisely written, 'No one can say why hearts will break and all marriages are opaque'.[12]

O'Connor's relationship with his father-in-law was probably not made any easier when the newlyweds occupied one of the houses built by Ness in Gloucester Street. But if Kathleen O'Connor's letter is any guide, her father had nothing to fear from William Ness:

> [There was] nothing [he] could have done to upset the work our father was doing as no one else could have undertaken it without great experience. Also he had all the Goring family as background— with the Goring and Yelverton families he had a wonderful beginning in New Zealand.[13]

O'Connor's connections may well have aided him to get started in New Zealand, as his daughter argued, but it was his innate ability, his engineering competence and his leadership that were more important factors contributing to his rise in government service. His name became publicly linked with major development projects and his contribution to the success of those projects was acknowledged—as it would be from now on throughout his career. In August 1874, for example, the extension of the railway line from Rakaia to Ashburton was opened with due ceremony, the extension completing the link between Christchurch and Ashburton, 56 miles south of the provincial city. Second in importance in the government party was the District Engineer, C. Y. O'Connor. At the celebratory breakfast held at the Somerset Hotel, Ashburton, and to which 'the early start from Christchurch added no little zest', the Secretary for Public Works toasted the health of the District Engineer in champagne. Mr O'Connor replied modestly, thanked the government minister, and referred to the advantages of the narrow-gauge principle, explaining that its necessary adoption had contributed to the delay in completion.[14]

O'Connor's sojourn as District Engineer in the big city, however, did not last long. Another promotion was on the way. Towards the end of 1874, the central government in Wellington appointed him District Engineer for the whole of the Westland Province and the southern part of Nelson Province. He and his wife and child (with another on the way) moved back to the West Coast to live in Hokitika.

The centralization and financing of development operations in the provinces had gradually weakened the power of the local councils since the late 1860s. By mid-1874, the North Island provinces were abolished and, despite fierce resistance, provincial assemblies in the South Island were abolished the following year. 'New settlers', wrote Pember Reeves

> with the ordinary British contempt for the institutions of a small community thought it ridiculous that a Colony with less than half a million people should want nine governments in addition to its central authority.[15]

O'Connor's responsibilities on his return to the West Coast were considerably broadened and more demanding than when he was there previously as assistant engineer and then District Engineer for Westland. Now he was set the task of improving the harbours of Greymouth, Westport and Hokitika, each with difficult sandbars; at Hokitika, the river estuary was so obstructed by shifting sand and travelling shingle that flooding was commonplace. He constructed improvements at these three ports and acted as consulting engineer to the local harbour boards. 'These [harbour works] were carried out successfully to completion with notable professional skill and judgement in the face of immeasurable difficulties.'[16]

The existing files for these years when O'Connor was District Engineer show almost daily correspondence with his chief in Dunedin, and with various authorities and contractors with whom he had professional business.[17] The

Greymouth Port, South Island, redesigned and constructed
by C. Y. O'Connor in the 1870s.
Courtesy West Coast Historical Museum, No. 012

volume of work, the variety, and the urgency of the many problems he faced
now seem prodigious. What emerges from a study of the papers is that
O'Connor, quite apart from his engineering capability, possessed remarkable
management skills: a capacity to understand and deal with a number of
demanding and often conflicting concerns at any one time. As an example,
within the space of one week in January 1878 he was working on repairs to the
Aranata Bridge; negotiating with a manufacturer for a locally constructed spoon
dredger; travelling to Greymouth and the Kumara goldfields to supervise the
building of a water race; arranging for the visit to the district of the Premier, Sir
George Grey; and, because all expenditure had to be authorized by the
Provincial Engineer, constantly submitting estimates, contracts and accounts to
Dunedin, and dealing with correspondence in return. Paperwork seemed
endless but there is no record of O'Connor's advice ever being rejected by the
distant Provincial Engineer in Dunedin.

In the files for January 1879, there is an indication of O'Connor's firm line
with contractors who sought revision or extension of their contracts. He
interpreted contracts strictly and was reluctant to compromise. He believed that
contractors were firmly bound by their original estimates and that these could
not be altered unless exceptional circumstances warranted a revision. A typical

[71]

letter to a supplicant contractor, dated 29 January 1879, expressed his views, often to be repeated:

> Having again carefully investigated the question, I have come to the conclusion that contractors, either in law or in equity, but especially in equity, are not entitled to any further payment, beyond what they have already got in the contract. I cannot therefore entertain the claim which you advance.[18]

He argued in this case, and in many others throughout his executive career, that unless contractors were made to work within their estimates, it would lead to false—that is, lower—estimates being submitted merely to obtain the contract, the contractor safe in the knowledge that the figure could be revised upwards at a later date. In another case in the following month, O'Connor wrote a strongly worded letter to the contractors Dixon and Edward Spier, complaining that the workmen's cottages they were erecting were unfinished at the time of expiry of the contract. He accused the company of not exercising 'due diligence' and required them to employ six extra men until the work was completed. 'If this isn't done and there is further cause for complaint, the Minister for Public Works may exercise powers vested in him.'[19] This strong resolve did not enhance O'Connor's reputation with erring contractors, but it earned him respect and trust from government departments, and gradually even from honest contractors who saw the justice in his argument.

O'Connor's keen sense of fairness in all his dealings could just as easily lead him to argue for compensation in cases where hardship had ensued from planning or building work. Confronted with the problem of nine squatters on land reserved for the Hokitika–Greymouth Railway, he recommended that the cost of moving five of them who were growing vegetables and flowers for market should be borne by the government and compensation paid in lieu of 'this year's crops'. For three of them who took up their land under business licences under the Goldfields Act, he recommended that they be treated as freeholders and allowed to claim full compensation.[20]

The provision of water supplies to the coal and goldmining communities throughout the province was also O'Connor's responsibility. A feature of goldmining on the West Coast at that time was the huge water races—aqueducts, usually supported by high wooden scaffolding—that brought much-needed water for sluicing purposes from the rivers and lakes higher in the mountains to the mines below. Often these water races, remarkable engineering achievements in themselves, would snake their way across the country, over valleys and hills, high above the ground, for a distance of several miles—13 miles in the case of the Waimea–Kumara field. Gravity provided the forceful heads of water capable of separating the gold grains from the sludge. Much of O'Connor's time and extant office correspondence was taken up with the building, maintenance and extension of these water races under the care of the government. The government had a heavy stake in the continued success of goldmining, and the District Engineer was at the beck and call of the mining companies.

As the gold deposits were mined out, rich coal deposits in the hinterland above Greymouth took their place. Mining 'black gold' became the main source of prosperity in the district. Coalmining also required water supplies and a safe, well-equipped harbour at Greymouth, which O'Connor designed and supervised, the extended stone breakwaters releasing the town from the constant flooding experienced until that time.

Thus roads, railways, water supplies, and constructing and maintaining harbours became the pattern of O'Connor's engineering experience—a pattern that would be repeated on a grander scale when he arrived in Western Australia.

Dillmanston water race at a West Coast goldmine of the type constructed and maintained by C. Y. O'Connor. Water was conveyed often over great distances down the mountains in the wooden aqueducts, increasing in force, to assist the processing operation at the mine below.

Courtesy West Coast Historical Museum, Shannon Collection, No. 1084

It was while O'Connor was at Hokitika that he first met the eminent English marine engineer Sir John Coode, who was retained by the central government to report and advise on the condition of all the harbours throughout New Zealand. Sir John had been knighted following his construction of Portland Harbour in the south of England, and before visiting New Zealand he had designed and supervised the construction of Colombo Harbour in Ceylon. Sir John Coode spent 17–19 April 1879 at Hokitika and inspected the marine work already undertaken by O'Connor. Back in London, Coode wrote his report using O'Connor's survey data and commended the quality of his work. Sir John's meeting with O'Connor, and knowledge of his capabilities, led him, when he was president of the Institution of Civil Engineers in London in 1880, to support O'Connor's membership of that prestigious institution.

Coincidentally, it was while on his way back from New Zealand to England that Sir John Coode stopped off at Fremantle to make his first inspection of the area and make recommendations for the development of a port in the vicinity. When O'Connor met him in Hokitika in 1879, there could have been no thought that O'Connor himself would one day play a far greater role in the building of that harbour and overturn Coode's recommendations.

9

When O'Connor returned to the West Coast as District Engineer at the beginning of 1875, he came as a married man with a family and set up house on Gibson Quay in Hokitika. Shortly after his arrival, he celebrated his 31st birthday; the last nine of those thirty-one years had been spent in New Zealand. He would have had reason to be proud of his progress from those early years as a trainee and fledgling engineer in Ireland and then as a raw immigrant surveyor uncertain of his next job, and his steady rise since then, both in responsibility and experience, in government service. District Engineer was a senior position and carried with it the then substantial salary of £700 a year with an allowance of £100, quaintly termed 'forage': an acknowledgment that as District Engineer O'Connor was required to keep a horse in order to travel around his district.[1]

Hokitika was then one of the busiest ports in the colony in terms of customs revenue and the number of vessels, both steam and sail, arriving from overseas. It was also one of the most dangerous: the harbour mouth was blocked by a treacherous sandbar negotiable only in certain weathers and at certain states of the tide. Vessels foundering, with resultant loss of life, were common occurrences. Although the harbour is no longer in use—all trading and fishing vessels use Greymouth, 25 miles north—there is at the harbour entrance a prominent beached hulk, a reconstruction of the schooner *Tambo*, one of forty-two ships wrecked at the harbour entrance. It is a permanent reminder of the town's costly maritime history. The beach hereabouts is disfigured with piles of logs, driftwood and flotsam washed down the Hokitika River in times of flood—grim evidence of the hazards to shipping in the past. The site of the O'Connors' house on the quay has been built over; it was within walking distance of the group of government buildings then in existence: the courthouse, customs, public works offices and police station.

We have some idea of O'Connor's character when he was appointed District Engineer from a profile of him by one of his trainee engineers, C. E. Bremner. O'Connor seemed, according to Bremner, abrupt in his manner and austere with his cadets, but he had a kindly and considerate nature and he unfailingly acknowledged service performed.[2] This accords with his daughter Kathleen's memory of him as sometimes fiery: 'He was known to dash his hat on the ground when angry and dance on it'.[3] But he was never violent. William Rolleston believed he 'personally combined gentleness and amiability with force and vigour of character and intellectual activity to an extraordinary degree'.[4] Bremner wrote:

> It was my privilege to enjoy his friendship and on Sundays to dine with him at his home. After dinner he would suggest a walk and at times this gentle exercise would extend ten miles before tea, with occasionally a small survey pegged off and with reference notes and sketch plan made.[5]

Bremner's reference to O'Connor's austere manner with his cadets was not overstated. He demanded of his trainees the same application, sense of duty and loyalty to the profession as he himself had displayed as a trainee. This is evidenced by his handling of a complaint made by his assistant engineer H. Gordon, who had charge of Nelsons Creek. Gordon had written to his chief about the poor conduct (unspecified) of a young cadet named Lundon—and it was evidently not the first complaint. O'Connor, showing that he is a stern boss who will not tolerate shoddy or poor work, writes back to Gordon, 'I should most assuredly on the spur of the moment have directed his instant dismissal'. But then, with that characteristic fairness for which he had become known, he gives Lundon another chance because 'so long a time has elapsed since the offence'. Nevertheless, he asks Gordon to deliver a strong reprimand and writes that had Lundon 'earned for himself a character for steadiness, willingness and subordination to his superior officers I should have had no hesitation in recommending him for the appointment'. (Lundon had been in line for an appointment as assistant surveyor on £250 per year, but in view of the complaints O'Connor gave the post to another.)[6]

As District Engineer, O'Connor's territory extended north and south of Hokitika and involved him in much travelling. He divided his time between three offices: his headquarters in Hokitika, which was a private house converted into offices; a branch office in Greymouth; and a third office in an old, disused public house in Ross, 15 miles south of Hokitika on the coast. According to Bremner, O'Connor had a staff of seven: a clerk, a draftsman, two cadets, two assistant engineers, and another engineer stationed at Greymouth. He was already a leader of men and an indefatigable worker who would keep his staff fully extended. He himself worked long hours.[7]

Within three months of arrival at Hokitika, Susan O'Connor gave birth to a second child, on 10 April. Christened George Francis but known in the family as Frank, he was destined, like his father, to become an engineer after the family's move to Western Australia.

Life for Susan O'Connor in Hokitika at that time could not have been easy; there would have been few comforts and social amusements. Her daily life in a raw mining town on the West Coast must have proved a painful shock, especially to someone of Susan's age and social expectations. She and her young family would have occupied a weatherboard house, primitive by modern standards—hot and dusty in the summer months, but wringing with damp, battered by rain and windswept throughout the long winter. Susan's home town, Christchurch, although hardly yet a metropolis, was fast developing into a bustling city with a lively social round; Hokitika, on the other hand, was little more than a shantytown, situated on the banks of the river of the same name, the streets awash with mud for most of the year. Melting waters from the glaciers, supplemented by the torrential rain, surged down the mountain sides carrying lava-like concretions of mud and piles of timber debris that wreaked destruction, sweeping away all obstacles in its path.

This problem of severe flooding and silting at the mouth of the river was the most pressing concern of the new District Engineer in the first months of his appointment. But Hokitika was not his only harbour concern: O'Connor also held in his brief the need for improvements at the mouth of the Grey River on which Greymouth was situated, and the mouth of the Buller River at Westport. These two, like Hokitika, were obstructed by difficult bars, which hampered the free flow of valuable timber and coal exports from the hinterland of the West Coast. O'Connor's strategy was to build substantial stone breakwaters at each port.

By this time, O'Connor was an experienced engineer, well equipped to deal with technical difficulties, but carping criticism, ignorant meddling and grandstanding by civic leaders were of a different order and it needed all O'Connor's humour and diplomacy to resolve them. On one occasion, in his role as engineer-adviser to the Hokitika Harbour Board, he reported on the condition of the river facings in the town and, going into elaborate detail, ended by recommending the renewal and restoration of the wooden piles. In a letter to the chairman of the Harbour Board, he complained about the chairman's letter published in the *West Coast Times*, which accused O'Connor of 'favoritism in awarding the contract'. He justified his action and added:

I may state that in continuing to act for the Board I do so on the faith of your promise of yesterday that all implication of favoritism on my part will be repudiated at the first opportunity.[8]

An example of O'Connor's humour in dealing with difficult civic busybodies is contained in a letter to his chief, William N. Blair, Engineer-in-Charge, Middle Island (as South Island was then called). O'Connor had recommended a certain course of action in covering the Kumara Water Race—a technical matter. Both the mayor and town clerk of Kumara had complained to Blair of O'Connor's recommendations, and made recommendations of their own. O'Connor defended his actions with accompanying reasons, and added in his letter to Blair:

Had Mr Seddon [the mayor] continued in office however, it would have been as well to give in at once however ridiculous it might have been to do so as he is proverbially one of those men, like Mr Jebson of Malvern in Canterbury, who stick to their points even if it is for years until they weary the other side into complying with their wishes.[9]

Memoranda and telegrams were exchanged daily between O'Connor on the West Coast and the Engineer-in-Charge, Middle Island, who was headquartered in Dunedin. Occasionally, the correspondence would refer to O'Connor's future benefactor, the road overseer Emanuel Rawlings—mainly requests for O'Connor to meet Rawlings for road inspections, or planning diversions on existing roads. His work ranged beyond harbour, road and railway planning. He oversaw plans for the Hokitika Lunatic Asylum, the lighthouse and the signal station, and negotiated with the local convent for resumption of its land, which at a later date, he was pleased to inform the trustees, was not required.[10]

While her husband was absorbed and often away from home directing and planning engineering works involving much travel along the West Coast, Mrs Susan O'Connor, like other pioneer women of her age and class, was left at home to manage the growing family. Her husband's salary and position in the public service made possible the employment of domestic help, although the shortage of young female settlers meant that most families existed without servants. Following the birth of Frank in 1875, Letitia Kathleen was born eighteen months later on 14 September 1877. Their third daughter, Eva, was born a year later.

At the time of Eva's birth, O'Connor was on one of his many visits to works north of Hokitika, in this case Kumara on the Taramakau River south of

Greymouth. In keeping with family custom, Susan wrote to her husband, asking what name the child should be called. He telegraphed back—jokingly, as it turned out—'Eva Dronghaia Hodderrena'. Small wonder that the registrar misspelt the names on the birth certificate.

It is an interesting glimpse into Susan's character that she did not see the joke, or want to see the joke, or query the accuracy of the telegram. She could hardly have been unfamiliar with her husband's bursts of humour. If she had doubts, she put them aside and interpreted the instructions literally when she could have easily waited until her husband's return. Eva was duly christened but, according to her older sister Kathleen, she 'never quite forgave her father'.[11] Understandably, perhaps, the young family, Scottish on their mother's side, did not always appreciate O'Connor's Irish sense of humour—humour that in this instance was enshrined on Eva's official documents ever afterwards. But there is much evidence to show that the children loved and respected their father and he loved them and spent as much time with them as his demanding job permitted.

With the birth of Eva, Susan O'Connor was faced with managing three daughters and one son all under the age of 5. It was not an unusual predicament in those days, but such toilsome family conditions could often result in high infant mortality and accidental death. Pioneer life in a small town was hazardous and particularly so for the children. In many cases, they were poorly supervised. Deaths by drowning, or from infant diseases inadequately treated, or from burns from oil lamps and cooking fires were common occurrences.[12] The O'Connor family was not immune from such tragedy. The brief life of their fifth child, born on 26 July 1879 and christened Charles Goring Yelverton, ended at the tender age of eight months due to scalding from boiling water. The accident may have occurred while Mr and Mrs O'Connor were at a reception held in O'Connor's honour in the Town Hall (see next page) on Saturday, 20 March 1880. The *West Coast Times* recorded on the following Monday that 'a severe and painful accident occurred to the youngest child of Mr O'Connor, District Engineer...a quantity of boiling water accidentally fell on the child, and it was severely burnt'. The child died on Monday, after the newspaper was published, and a death notice appeared the following day.[13] One family story recounts that the infant Charles was accidentally bathed in boiling water; a somewhat less recriminatory but no less horrific explanation suggests that the little child grabbed the handle of a pot of boiling water on the hob. But at the age of eight months (the inquest record states, erroneously, seven months), it would seem unlikely to have been self-inflicted. The coroner, Dr Giles, who attended the dying infant at Gibson Quay, states baldly: 'Death by scalding which was accidental'.[14]

Whatever the circumstances—or blame to be apportioned thereon—there is no mention of the tragedy in the O'Connor papers. To lose only one child out of eight may even be counted fortunate in an age when all graveyards carried a high proportion of headstones bearing testimony to pitiful infant death. Today there is not even a headstone for baby Charles O'Connor. The position of his grave in that part of the hilltop cemetery overlooking the Tasman Sea is now the cemetery car park.

O'Connor's second sojourn on the West Coast lasted only six years until 1880, when, at the age of 37, he received another promotion and an extension of his responsibilities. By this time, the Department of Works had been reorganized and new staff appointments had been made. John Blackett was Engineer-in-Chief; William Newsome Blair, Engineer-in-Charge, Middle Island. O'Connor became Inspecting Engineer for the whole of Middle Island. The departmental headquarters were located in Dunedin, then the largest and longest established city and capital of the Otago Province. The O'Connors and their four children now faced another move, a tiring, uncomfortable journey by coach back along the road to Christchurch that O'Connor had surveyed. Susan O'Connor and the children would surely have been reunited with her parents, who were still living in Avonside, Christchurch. Her husband is likely to have preceded them to Dunedin to establish a new home for them in that southern city, a city Scottish in character and population.

The news of C. Y. O'Connor's impending departure from Hokitika seems to have been universally regretted, as evidenced by the extravagant but doubtless sincere praise given him at a farewell reception at the Town Hall on Saturday, 20 March 1880. About forty leading citizens were present including the mayor of Greymouth and the Honourable J. A. Bonar, deputy mayor of Hokitika, chairman of the proceedings in the absence of the mayor. O'Connor was presented with a silver tea service, a clock and a diamond ring, and Bonar suggested that 'the last named Mr O'Connor wear in remembrance of his friends'.[15] Bonar spoke of O'Connor's 'sterling qualities', his 'many friends', and how 'he had done his duty thoroughly in the department over which he had control and he never spared himself in the execution of the most onerous work'. Bonar spoke of O'Connor's courtesy, and made a reference to the recent 'particular circumstances with public works on the coast which made it very trying for Mr O'Connor, who nevertheless, had done his duty with satisfaction to himself, the government, and the public'. This was a reference to a bitter strike by fifty-eight men employed on the Greymouth Harbour works back in January of that year. The two local papers and the mayors of the two towns had supported the men's action, which was taken in protest at a reduction of their wages from 10 shillings to 9 shillings per day. O'Connor, as the district officer in charge, was appealed to. In his reply, he refused the men's claim and supported the wage reduction, stating that 'wages on the west coast are immensely in excess of anywhere else in New Zealand'.[16] But he was clearly under instructions to do so by the government, a fact not missed by the local newspaper when writing about the affair: 'Our own District Engineer—who is recognised universally as a man of ability and experience—is bound hand and foot'.[17]

O'Connor's speech in reply at his civic reception is reported in full in the *West Coast Times* of Monday, 22 March. It is singularly revealing of his style, being one of very few records we have of O'Connor speaking about himself, his ideas and his experience. For this reason, it is worth quoting at some length. A statement in the same issue alongside the report of the speech confirms that O'Connor himself believed the transcript accurate, 'with the exception of the beginning and concluding sentences' (not quoted here):

That I should allude to so great a theme [the mining industry] upon a purely personal occasion may at first sight seem strange, but looking at the matter in all candour it ceases to be strange, for it is to the existence of mining industries here that I owe the boon of your acquaintance and friendship, and it is to the fact of our having been associated for many years in the hearty endeavour to promote such industries that I owe your kindness to me today. It may be excusable then, if I should endeavour to show you, before we part, that I have not lived in the midst of a great mining community for fourteen years entirely with my eyes shut; and it is even due to you, and the enlightening influences of the peculiar social conditions existing here, to show that I have reaped from the study of them some little of the great principles of living humanity which they are so well calculated to teach. To tell this plain unvarnished truth indeed, I may say at once, that such as I am, be it good or bad, I am the simple resultant of surrounding circumstances in which I have lived in the fourteen years of my life that I am conscious of having been a deliberating and thinking being, and this fourteen years it is almost unnecessary to tell you has been spent almost entirely upon the goldfields. Arriving here from home, a mere half-educated boy, saturated in old-world formulae, and representing merely the outcome of the thoughts of others impressed upon the elastic mind of a child, I found myself suddenly launched into a community of original thinkers, where each man was an individual, and where, at any rate, every tenth man one met with, thought entirely differently from oneself on all the debate-able subjects on earth and under heaven. I would ask then can anyone fail under such circumstances to think a little? Can anyone fail, having thought a little, to find that the conservative prejudices of the one particular school in which he was brought up are not all right, and beliefs and thoughts and practices which he had been carefully trained to look upon with horror, were believed in and thought and practised by men infinitely wiser and better in every way than oneself? Or, in a word, can anyone fail, under such circum-stances, to become less of a conceited, exclusive, and pedantic boy, and more of the hearty cosmopolitan, more of the liberal-minded sympathiser with all created beings? I think not.

O'Connor then, in what sounds like a modern defence of the rich benefits of multiculturalism, draws parallels between the North American gold rush immigration experience and that of the Australian colonies—a term he uses to include New Zealand:

The rate of progress of the American immigrants in their adopted country was most startling, not alone in agricultural pursuits but also in all arts and in all social institutions in which creative intellect could be brought into play. It dawned upon thinking men to consider the exhilarating effect of the admixture of races, each race bringing with it into a free country the thoughts that hitherto had run in definite grooves and then co-mingling with all other races in one harmonious effort to progress. This I take it is the distinguishing characteristic of the goldfields of the Australian colonies. Free as the air to all nationalities we have here assembled the countrymen of Socrates and Plato, Caesar and Galileo, Goethe, Schiller, Shakes-peare, Burns, Stevenson and Burke. All met together, and while each is preserving, to some extent, patriotism of the particular nationality from which he sprang, all are dying to outdo the others in a grand march towards a stupendous goal of progress to which the oppor-tunities of any country or nation have ever tended.

O'Connor then turns his attention to the system of education current in New Zealand at that time and expresses concern for what he sees as the uniformity and lack of variation in teaching:

I fear that [this] tendency [will result in] a nation without an idea of interchanging with each other, and moving so emphatically in the same groove that all individuality will be lost. What I think is the most important object of education is to create thinkers and reasoners, rather than mechanical beings trained to start at the sound of a whistle, and revolve round a common axle. I believe diversities of thought, instinct and habit, are the soul of motion in living beings.

O'Connor concludes his speech by reaching back into his Irish experience and turning his attention to the need for more poetry, ballads and folk stories inspired by the beauty, the grandeur and the calm of the mighty mountain ranges, the awful majesty of the banked-up storm, and clouds of rain—all experienced continually on the West Coast. 'Up to the present we have no such home songs', he deplores, 'and although the songs of our fatherland may do for us and for our generation, they will speak to our children more or less in a foreign tongue'. Here O'Connor shows himself to be sensibly committed to his new country—forward thinking, and not hanging on to the past or, like many of his countrymen at that time, revelling in nostalgia for an idealized Ireland.

He finishes by thanking all those who have been his friends and colleagues during his stay on the West Coast, and the assembled group for the handsome presents and the welcome he had received that evening. 'I leave you in wishing all possible prosperity, good luck, and God speed to the mining industries of the Australian colonies, and especially to Westland'.

This was the second occasion on which he had been honoured for his work in Westland, and the second such presentation that had been made to him by leading citizens in recognition of his work.

The speech provides us with a unique insight into the character of O'Connor. He reveals himself to be a moralist, an intellectual, and not solely a technically competent man. There is much that is very modern about his thinking and much of it in harmony with the ideals of Australia and New Zealand today; his thoughts on education, multiculturalism and the need to develop a national response to the arts would not disgrace a modern politician. Unique the speech might be in the sense that no other extant O'Connor document is so revealing of character, but the ideas expressed were not unique among cultured engineer-idealists of the Victorian era. For them, the building of bridges and harbours and railways was not motivated solely by thoughts of economic gain but was part of a triumphant progress towards a better world—a moral, cultured, prosperous world for all.

That weekend in Hokitika was not all celebration. Sadly, in the same issue of the *West Coast Times* in which the speech was printed, appeared the news of baby Charles O'Connor's accident, which resulted in his death.

Oddly as it may seem after all his years of practical experience, O'Connor still had no formal engineering qualifications. At this point, he sought to regularize his professional standing by sitting an examination consisting of Algebra and Trigonometry papers, supplemented by an oral examination and submitting evidence of his nineteen years in the field. Not surprisingly, he was confirmed as a surveyor 'qualified to undertake surveys in accordance with the system in operation in this colony'.[18] In all areas, save one, his results were classed as 'very good'. Only the standard of his 'penmanship' failed to receive a similar accolade from the examiners. (Anyone perusing his diaries and letters would heartily agree.) O'Connor also received, in March 1880, confirmation that he had been admitted as a Member of the Institution of Civil Engineers in London, his membership proposed by his old mentor, John Chaloner Smith, seconded by his engineer superiors in New Zealand, and endorsed by the president of the institution, Sir John Coode. In later years in Western Australia, when he was attacked in some newspaper columns for being underqualified— 'this shire engineer from New Zealand'—he was able to defend himself adequately.[19]

O'Connor's life in Dunedin evidently provided him with increased opportunity for indulging his favourite recreations. He became a member of the Otago Hunt Club and attended meetings. One of his cadet engineers remembered him 'flying over fences and stone walls on his big chestnut with his white mackintosh coat, generally unbuttoned, streaming over the horse's tail'.[20]

When O'Connor took up his appointment as Inspecting Engineer for Middle Island, the railway line linking Christchurch with the West Coast had yet to be constructed in spite of public pressure demanding this important facility. One of his first duties on appointment was to give evidence to the Railway Commission appointed to investigate and decide on the best route for the line to follow. O'Connor had extensive personal knowledge of Arthurs Pass and was able to support his choice of this route with his statistics and his 'lucid and comprehensive' special knowledge of the terrain.[21] Although the Railway Commission was satisfied, the government was not convinced and ordered further investigations of other routes. In each case, the commission inquiries confirmed O'Connor's claim that Arthurs Pass was the best route. And this was where the railway was eventually constructed.

After three years as Inspecting Engineer, Middle Island, O'Connor's total service in New Zealand amounted to eighteen years. He was then aged 40 and felt much in need of long-service leave, to which he would have been entitled. An earlier application for overseas leave had already been approved. This, however, was to be rescinded with the offer of another government appointment that would radically alter his career, his living arrangements and his status.

It was less an offer than a royal command that arrived by surprise telegram from cabinet minister William Rolleston, appointing O'Connor Under Secretary for Public Works based at the seat of government, Wellington. This was soon followed by an explanatory letter, dated 19 November 1883, which enclosed the warrant signed by the Governor, William Francis Drummond. Rolleston's letter began by stating the government's appreciation of the sacrifices O'Connor would make in accepting the appointment, notably his leaving the practical work of civil engineering, 'a profession in which you have rendered such long and efficient service to the colony'.[22] Rolleston then gave an assurance that if O'Connor chose to return to the Engineering Department of the government in the future, his position and claims to promotion in relation to other officers in the department would not be affected by his period of absence as Under Secretary.

> With regard to what was said with regard to the leave of absence which you contemplated taking, it would not be possible for the present to give effect to what was intended consistently with acceptance of this offer.

Rolleston's letter ended by requesting that O'Connor take up his duties in Wellington by 25 November, and confirmed that the salary would be £800 per annum, with a special allowance of £200 to cover the cost of removal and travel to Wellington.

The young family must by this time have been attuned to the routine of moving house. With this latest move, the prospect of living in a comparatively large city, with all its conveniences and rich social life, must have seemed like an exciting change—and possibly it made up for the disappointment of delaying indefinitely the planned overseas holiday. At the time, Aileen was 10 years old, Frank 9, Kathleen 8 and Eva 6. The baby of the family, 2-year-old Roderick, had been born in 1881. Wellington would be their home town for the next seven years and there they would experience school and new friendships. They would be close to their New Zealand cousins, and their father's important new job would mean less travel and a higher salary, and would provide Mrs Susan O'Connor with a satisfying social round.

IO

WHETHER TOWARDS THE close of 1883 C. Y. O'Connor had any misgivings about accepting his new government appointment in Wellington we cannot be sure. But from what is known of his cautious delay in accepting Sir John Forrest's offer seven years later, the change would surely have caused him much anguish. True, there was prestige attached to his new post, and a slightly higher salary, but he was exchanging the practical world of engineering, for which he had been trained, for a largely administrative post, a life tied to an office desk as a secretary in the service of his head of department, the Minister for Public Works. His previous work as District and then Inspecting Engineer had required administration but it was closely identified with his fieldwork; now, in Wellington, the administration would be divorced entirely from the field.

Apprehensive or not, O'Connor had little choice but to accept. And it was surely natural for him to have felt proud of the recognition given him. He was a man who quietly but firmly knew his own worth and was confident of his abilities.

Henceforward O'Connor would be administering other engineers, arguing on their behalf for government appropriations, writing his minister's speeches, checking and questioning tenders and advising his minister on priorities—but he would be unable to leave his mark in a practical way on railway, harbour, mining and road constructions.

He would have been aware, also, that he was taking up a government administrative position at a time of increasing economic hardship in New Zealand. The loans secured by Julius Vogel in the early 1870s had been spent; a new, more frugal public works and immigration policy had been adopted. Prices for exports were tumbling, the price of land had plummeted, private investment had declined—all resulting in a prolonged period of unprecedented economic depression. Retrenchment and unemployment were widespread, particularly in the Canterbury and Otago provinces. Hundreds of men, some of whom may previously have worked on O'Connor's engineering projects, could now barely exist on government relief work, which paid a subsistence wage. That O'Connor had great sympathy for them in their plight is indicated by the handwritten draft of a letter (undated) that he wrote to the *New Zealand Times*, defending his department's decision to grant a holiday on the occasion of Canterbury's Foundation Day. The paper had questioned the wisdom of the department 'agreeing to pay wages to the unemployed in Christchurch during a day's holiday'. The newspaper's attack centred on O'Connor, who was, it argued, a previous Canterbury resident and had therefore favoured the people there. In his reply, the recently appointed Under Secretary reminds readers that the day in question

> is always kept as a special Festival Day. The desire of the people of Christchurch to give the unemployed a holiday and a dinner on that day is due chiefly to the fact that they could not comfortably take the holiday themselves while a number of their fellow colonists in their midst were precluded from joining with them through no fault of their own.

He ends his letter with a barb aimed at the more affluent people of Wellington:

> I think we should be in very great straits indeed before we would be justified in denying one holiday in the year to the unfortunate ones amongst us…and I am sure that anyone who thinks about it at all and realises the number of holidays enjoyed by the community in Wellington while receiving full pay, will admit that there is nothing either lavish or unprecedented in granting one holiday in the year to such of our fellow colonists as can only procure employment on the relief works.[1]

Here, in the concern shown by O'Connor for the unemployed, is surely an echo heard forty years previously by a small boy in the old house, 'Gravelmount', at a time when his father was showing similar concern for the Irish unemployed on relief during the Famine. O'Connor, the son, carried with him throughout his life a compassion for the working class that owes its origin to his father's influence and his formative evangelical upbringing.

It is not likely that any misgivings felt by C. Y. O'Connor on his move to Wellington would have been shared with his family. In households in Victorian times, the husband invariably made the decisions relating to all family and business matters. As heartless as this may seem to us today, the conscientious Victorian husband would have seen it very differently. It was his moral duty, he believed, to shoulder all the worries and responsibilities that weighed upon him, thus sparing his family an anxiety that could be avoided. Therefore, similar misgivings would not have troubled Mrs O'Connor and her four children as they settled into their new home, a double-storeyed timber house in Tinakori Road, a long straight street on the rise, overlooking the harbour and in a fashionable quarter of the city. In modern times, the street overlooks the Wellington Urban Motorway and is chiefly visited by tourists seeking hotels and Katherine Mansfield's birthplace. The O'Connors' sixth surviving child, Bridget, known in the family as Biddy, was born the year after the family arrived in Wellington. Completing the family, another son was to follow: Murtagh Yelverton Goring.

The two elder girls, Aileen, aged 10, and Kathleen, aged 7, were almost immediately enrolled at Mrs Swainson's School, an Academy for Young Ladies, in Fitzherbert Terrace. Later known as Marsden Collegiate School, its most famous pupil in subsequent years was Katherine Beauchamp, better known as Katherine Mansfield. Eva, who was only 5 when the O'Connors moved to Wellington, probably joined her sisters at the school at a later date. Kathleen's closest childhood friend, Estelle Beere, whom she met at the school, visited the O'Connor house every Saturday. The girls travelled by horse-drawn tram, and on Sundays Estelle went to church with the O'Connor family. O'Connor would always give the children threepences for the collection plate and Estelle remembered many years later, when writing to Kathleen O'Connor, how 'He always gave me just as he did you [Kathleen] and Eva. I look back on his sweet kindness to his children's friends'.[2]

The tasks that O'Connor faced as Under Secretary were of the order of Hercules'. The pressures upon him, coming both from the government and the engineering departments that he administered, to maintain an acceptable level of public works in a period of economic depression, falling budgets and retrenchments would have daunted a lesser man. That he overcame these diffi-culties to a large extent is indicative of his capacity for hard work, his organizational ability and his grasp of the subject before him in all its detail. From the official records, including his inter-departmental letters written at that time, it is clear that he dealt successfully with an enormous volume of work. But in spite of this heavy workload and responsibility, his income as Under Secretary was, in fact, less than he had received as District Engineer, if his supplementary consulting fees from the Harbour Boards of Hokitika and Greymouth are taken

The O'Connor family in Wellington, c. 1891, just prior to departure
for Western Australia.

Courtesy Battye Library 67787P

into account. For this reason, he fought hard for an increase in salary of £200,
'on account of extra work in connection with various reorganisations and
reforms'.[3] The work was so excessive, and the salary so inadequate, that he
petitioned the government to be restored to the Engineering Department. In
supporting his claim, the Treasurer, Julius Vogel, acknowledged that 'O'Connor
is entitled to great credit and thanks for the remarkable improvements that he
has made'.[4] But parliament would not authorize a permanent increase, preferring
to pay O'Connor a bonus, which he had to re-claim and justify each year.
Neither would they agree to him returning to the Engineering Department.

Although the quality of his work in the department was never to suffer, it
was clear at an early stage that his employment in New Zealand Government
service was far from a rewarding time for O'Connor. His sense of injustice and
uncertainty grew, he felt that he was not being treated fairly, and as a result it
was not long before he began to consider opportunities for engineers overseas.
His former engineer colleague William Smith was one of those urging him to go
to New South Wales, 'the best place in the colony at present and for many
years'.[5] Smith advised him to get leave, go to Australia and look for himself. 'No
one will be any wiser of your visit here', he adds. Another correspondent,

William Evans, wrote from Sydney hoping to recruit O'Connor as a business partner in building railways in New South Wales.[6] These enthusiastic letters from Australia must have tempted O'Connor to some degree, coming as they did soon after a reduction in his salary in line with other stringent government economies. The outlook was bleak, but in the matter of leaving one colony for another O'Connor was cautious. His young family considered themselves New Zealanders and they were happy in their life in Wellington. Although his daughter Kathleen was later to write that her father 'carried Ireland with him everywhere', there is no evidence that he yearned to return there. He had made New Zealand his home; it was the country he had known for most of his adult life, and when he eventually came to leave he did so evidently with regret, writing, 'I should certainly not have left if I could have got security'.[7]

In his first five years in Wellington, 1884–89, leaving the country was never more than a fleeting option. His day-to-day work demanded his full attention, concerned as it was with administering and overseeing government policy relating to public works. A pressing task each year was the compilation, writing and proof-correcting of the minister's annual report to parliament, outlining past achievements and future public works policies based on information supplied from each of the engineering sections—roads, railways, harbours and mining. Each successive May in the years he occupied the post, he would laboriously draft, in long-hand, letters to each of the secretaries of the different departments, asking

> if you would kindly supply a short memo containing the substance of what the minister for Lands [or Railways, or Mining, or Harbours] would wish to have said in the Public Works Statements, and also your proposals for the future.[8]

For a man who himself had experience of planning and building harbours, roads, railways and bridges, this must have been dull work indeed. Reading these reports at this distance, it is sometimes possible to see O'Connor's clarity of thinking and his organizational ability behind some of the proposals put forward by his minister—the appointment and selection of boards to run the new railways as they came into service, for example, or the minister's refusal to accept twenty new engines ordered from England that were found to be 10 tons heavier than specified—so heavy in fact that 'it would have been necessary to strengthen all bridges on the lines they were to run on'.[9] The engines were returned and replacements swiftly imported from the United States. In another report on the North Island trunk line, presented in parliament in 1886, the draft, in O'Connor's hand, argued that 'When the votes were cut down last year, several of our survey practices were broken up and consequently the work of finally locating the line has been much interfered with'.[10]

In these desk-bound years, O'Connor must have been sustained by the promise, originally made by Cabinet Minister William Rolleston in his letter of appointment, that O'Connor might return to the Engineering Department in the future, and that his seniority in that department would not be affected by his service as Under Secretary.[11] Also, his spirits must surely have been raised by the entirely unexpected legacy left to him by Emanuel Rawlings. He would have known of Rawlings's tragic death in 1887, but the will leaving O'Connor £3,000 was not discovered until a year later. Such a sum, at that time without death duties, would have enhanced considerably O'Connor's financial standing.

Such hope as O'Connor had of returning to engineering must have received a boost when his friend and one-time fellow engineer John Blackett resigned as Engineer-in-Chief to take up the new colonial appointment of Consultant Engineer in London. He left New Zealand in February 1889 on a salary of £800 a year, but not before he had written a generous testimonial for O'Connor, an indication perhaps that both he and O'Connor expected the latter to apply for the position and be appointed as Engineer-in-Chief in Blackett's place. He concluded his summary of O'Connor's experience by writing:

> I have a high opinion of your ability and skill as an engineer and could with confidence recommend you for employment in any kind of work that might offer which demanded for its successful carrying out a good administrative ability, combined with untiring industry and application.[12]

But the government was in no hurry to fill the vacancy, relying for the time being on Blackett's assistant, W. N. Blair, who was then in poor health. The delay in making an appointment provoked criticism of the government in the newspapers. Opinion was divided on who should succeed Blackett in this important post, Blair or O'Connor. The *Wellington Evening Post* strongly backed the bluff Scot William Blair, and imputed to O'Connor an inadequacy for the job. The *Grey River Argus* in turn attacked the *Evening Post* and claimed that O'Connor was the best man to succeed Blackett. In praising O'Connor, the paper makes a veiled reference to what it saw as a weakness in Blair's character. It opined that O'Connor is by no means the kind of trimmer who spreads his sails to catch the breath of applause. When his judgement tells him that he is right he has a habit of sticking to his opinion.[13]

Yet another newspaper reported that a deputation 'waited upon Mr Guinness MHR to urge him to use his influence for getting Mr O'Connor appointed to the position of Engineer-in-Chief, vice Mr John Blackett'.[14] Extracts like these show that O'Connor, by 1889, was a well-known public figure in New Zealand and that his worth was recognized and appreciated, if not universally then certainly by the majority of influential citizens.

When the appointment to fill Blackett's vacancy was finally made in May 1890, O'Connor was not only disappointed with the government's decision but forced to reconsider his future in New Zealand. In continuance of its policy of retrenchment and austerity, a plan was implemented to combine the positions of Engineer-in-Chief and Under Secretary. The then Minister for Works, Thomas Fergus, confirmed in a letter to O'Connor that Blair had been appointed to this joint position. The letter continued:

> I have now to offer you the position of Marine Engineer for the colony with special control of Westport and Greymouth...the conditions under which you hold the appointment will be in all respects similar to those under which you now work.[15]

By accepting the post of Marine Engineer, which he did on 12 May, O'Connor had his wish to return to the practice of his profession but it was tempered by the knowledge that it was a lesser appointment—more a consolation prize than one befitting his true worth. This was partly acknowledged in Fergus's letter of appointment in which he added:

> in the event of Mr Blackett's retiring from the position of Consulting Engineer for the Colony, a contingency not impossible at an early date, it is intended to offer you a position on the same terms and conditions that he now enjoys.[16]

Tauman quotes a colleague of O'Connor's recalling that it was the only occasion that he saw him distressed and agitated.[17] He was so agitated, in fact, that in accepting the post, he asked that it be distinctly understood and agreed on behalf of all concerned that the appointment of Marine Engineer was an independent one and not subject to the supervision of Blair.[18] O'Connor had little respect for Blair's professional competence. Adding fuel to O'Connor's sense of injustice, Blair continued to be absent from his office, a sick man, and the Marine Engineer was left with much of the work of Engineer-in-Chief, unpaid and unacknowledged.

The minister attempted to allay O'Connor's concerns by confirming that his independence would be officially recognized, and repeated his earlier statement that had the government not appointed Blair to the position of Engineer-in-Chief, they would have offered the position to O'Connor.[19] It was a strange, clumsy excuse that did nothing to assuage O'Connor's growing discontent.

The seeds were now sown and germinating for O'Connor's imminent departure from New Zealand.

II

———

ALTHOUGH O'CONNOR'S LAST year of service in New Zealand was not a particularly happy one, beset as it was by nagging uncertainty about status and permanency, compounded by the reluctance of succeeding governments to give him assurances, there were compensations. Principal among these was that he was returning to his profession, which involved him once again in day-to-day practical engineering problems in the field—the field, in this case, being harbour works throughout the entire country.

Safe harbours were particularly important in the developing economy of New Zealand, a maritime country that relied heavily on sea transport not only for world trade but for communication within the country itself. The railway network was still in its infancy at that time, and the cheap transportation of goods and passengers from one inaccessible part of the country to another was largely achieved by sea.

The work of Marine Engineer freed O'Connor from constant desk work, requiring him to travel to and from ports in the colony, advising, recommending, reporting. One of his early journeys was to the port of Gisborne on the east coast of North Island in August 1890. Sir John Coode had originally designed the harbour at the mouth of the Gisborne River in 1880, but the design and location had been changed by the local Harbour Board, resulting in a totally inadequate facility threatened by sand travel. O'Connor recommended modifications to rectify the mistakes already made, and in doing so showed how much knowledge he had acquired on the vexed subject of sand travel in river harbours—knowledge that would be crucial in defending his designs for Fremantle Harbour at a later date. Sand travel was, he explained

> not merely a local shifting of a limited quantity of sand, but a
> continuously progressive travel of sand in one direction for a long

distance from a practically inexhaustible source. If the latter were the case at Gisborne the mischief would I conceive be incurable by any expenditure within reason.[1]

Another lengthy journey he made in company with engineer John Goodall was to Timaru, a harbour town on the South Island, south of Christchurch. There the problems were much the same. Advice from Coode and others twenty years earlier had again been rejected by previous unqualified officials. A later Harbour Board, in place in 1890, had asked for help from the Marine Engineer. O'Connor visited them and reported that he considered danger was imminent not from sand, in this case, but from travelling shingle. He recommended dredging to control the deposits, the extension of a mole and the building of groins. He also recommended the best dredging equipment to use, the methods to be employed, the costs and the frequency of dredging. Later, after submitting his detailed report to the board through his minister, he received a letter of appreciation from the Timaru Harbour Board commending him on 'the skill, care and attention' that he had devoted to their problems.[2]

The practical work of Marine Engineer must have delighted O'Connor, although early on in his engagement there were disquieting signs that his future in New Zealand was insecure. The health of the incumbent Engineer-in-Chief, W. N. Blair, had continued to deteriorate, and to relieve the burden on O'Connor the minister appointed William H. Hales as Assistant Chief Engineer. Hales was thirteen years older than O'Connor, and yet was someone with less experience. Once again, the government had shown its unwillingness to recognize O'Connor as the natural successor to first Blackett and then Blair. Adding to this disappointment, O'Connor's second request for leave was rejected, and he looked on with frustration and a sense of injustice as able men in the engineering and surveying departments were retrenched.

Looking back on that year 1890–91 from this distance, we can chart O'Connor's growing disillusionment and the inevitable break with New Zealand as a logical and foreseeable development. But, as in some Greek drama unfolding slowly, the resolution is seldom clear to the hero until the last act. O'Connor must have wondered many times how and from what direction his fortunes might change; his anxiety about his future caused him grave concern. We know that in this period he considered openings available overseas but none was sufficiently promising to gain his enthusiasm.[3]

Whether O'Connor took any interest in the National Australasian Convention, which met in Sydney in March 1891, we do not know. The convention, deliberating in the Sydney Parliament Assembly building, and lasting six weeks, came to no firm conclusion on the proposals for federation, but it happened to bring the Western Australian Premier, John Forrest, in contact with two delegates from New Zealand: recently defeated Labour Premier, Sir Harry Atkinson, and one of his former cabinet ministers, Captain William Russell. As remote as the convention's deliberations may have been to O'Connor, it was the meeting between Forrest and the two New Zealand delegates that provided the opportunity for a dramatic new turn in O'Connor's career and his release from his deteriorating relationship with the New Zealand public service.

At the time of their meeting, Western Australia had been an independent, self-governing colony for only three months. John Forrest, the first Premier, was a forceful, genial visionary, a towering figure both in size and influence, who was convinced that only a bold policy of extensive public works, financed by overseas loans, would lift the colony from its previous mendicant position to become viable and prosperous. However, just as Napoleon III needed a Haussmann to build modern Paris, Forrest, to realize his objectives, needed an engineer partner, a practical man as committed, talented and visionary as himself to take those ideas and convert them into realities. Therefore, his search for a Chief Engineer for the newly independent colony had become one of Forrest's priorities in his first months of taking up office until his departure for the convention. And he believed, for a while anyway, that he had found the suitable candidate: H. S. Mais, formerly Engineer-in-Chief in South Australia, and afterwards in private practice in Melbourne. But when Mais was offered the post at £1,250 a year, he declined, stating his preference to continue enjoying the rewards and satisfaction of independent practice. When Forrest arrived in Sydney on 9 March, having missed the first three days of the convention, the problem of finding a Chief Engineer had not been resolved.

The seven Western Australian delegates joined 'the most impressive gathering of politicians ever to have assembled in Australia':[4] six Premiers, including Forrest, nine former Premiers, cabinet ministers and former ministers, and several other representatives of Australian and New Zealand parliaments. And like any gathering of this kind, what was not achieved in the debating chamber was made up for in social meetings and discussions in a less formal setting. It was in these circumstances that Forrest unburdened himself to Atkinson and Russell about his dilemma of finding a suitable Engineer-in-Chief.

Another key figure in John Forrest's party was O'Connor's 1860s regional supervisor on the West Coast, his old friend and mentor, Malcolm Fraser, now Sir Malcolm Fraser and shortly to take up his appointment as Western Australia's first Agent-General in London. After leaving the West Coast of New

Zealand, Fraser had, in 1870, been appointed Surveyor-General of Western Australia, where he had reorganized and modernized the Lands and Surveys Department, making John Forrest his deputy. He was a member of the Executive and Legislative Councils under Governor Broome prior to Western Australia becoming an independent colony, and later was Colonial Secretary. Fraser had evidently not forgotten his old New Zealand friend and staff member, because it was he who signed the cable to O'Connor advising him of the vacancy, which ran as follows: 'Engineer-in-Chief W.A. Vacant. Salary not under £1000 per year. Cable application John Forrest before Monday, Melbourne. Self and Atkinson have recommended you strongly'.⁵ This suggests that not only Fraser and Russell but also the former Premier of New Zealand, Sir Harry Atkinson, knew of and sympathized with O'Connor's treatment, to the extent of helping him to find alternative employment. Since Atkinson had recently been forced to resign and Russell was then a member of the Opposition in New Zealand's new parliament, their sympathy for O'Connor under a regime that they opposed may have added to their conviction that a man of O'Connor's abilities deserved to be rescued from a beleaguered public service.

Fraser's cryptic message arrived in O'Connor's office in Wellington late in the afternoon of 8 April 1891 after the inconclusive convention had broken up and when Forrest was preparing to leave for Perth. It was the start of a month-long series of almost daily messages and letters exchanged between O'Connor and Forrest, and a sequence of memoranda between O'Connor and his minister, and the new New Zealand Premier, John Ballance—some of them recriminatory—which culminated in O'Connor's severing his ties with the public service at the end of the month.

Fraser's cable goaded O'Connor into action. Within twenty-four hours of receiving the message, he replied indicating his interest in a cable to John Forrest in Melbourne: 'Understanding vacancy Engineer-in-Chief WA salary not under £1000 five years guaranteed have honour to make application for same. Record of service posted today addressed you in Perth'.⁶ On the same day, he wrote to Forrest with a fuller application and explained that he had only heard of the vacancy the previous evening. Because the mail was leaving that day, he had only time to enclose the briefest testimonials and letters of commendation from Blackett and Vogel.

Probably Forrest felt that O'Connor had no need of extensive testimonials. What the New Zealanders and Sir Malcolm Fraser were able to tell him about O'Connor was merely confirmation of his suitability. Formal confirmation was received later from the Commissioner of the New Zealand Government Railways, James McKerrow, who wrote privately to Forrest:

His skill as a practical and scientific engineer is undoubted. He is an unwearied worker and at the age of 48 is in prime possession of his powers. Mr O'Connor's probity and integrity of character are beyond question. He is of very good presence and address and while firm in purpose has the knack of getting on well with those he has to deal with...I think it would be a mistake for our Government to let him go.[7]

John Forrest, in his dealings with people, was 'sincere, humane, trusting and honest'[8] and in this case he knew he had found his man. But 'his man' was cautious and would not move until he had assurances about pay and conditions. Forrest responded to O'Connor's first cable immediately prior to his departure from Melbourne, asking the earliest date that the engineer could take up duties in Western Australia. In Forrest's mind, at least, the matter appeared settled. But O'Connor hedged in his reply of 11 April, saying that he had to give three months' notice, although he thought the government would probably waive this requirement. He added in his cable that he had written to Forrest on 10 April on the subject of salary, addressed to Perth, and he would like Forrest to read the letter before making a decision on this important matter.

There followed another exchange of cables on 13 April in such quick succession that often one cable does not appear to be answering another. The first was sent from Forrest firmly offering O'Connor £1,000 a year and expressing a willingness to facilitate arrangements for O'Connor's release from the New Zealand Government. O'Connor's cable on that day may have crossed with Forrest's, because he appears not to have known the salary offered. He asks the extent of work expected of him, whether railways only, or harbours and roads. He thought the salary should depend on the extent of work and reminded Forrest that he would be giving up a pension in New Zealand. 'If offer £1500 would hesitate no longer. If not £1500 kindly say maximum and work involved.' Another cable from Forrest on that day is merely a postscript stating that he is leaving for Adelaide. And then from Adelaide on 14 April comes the answer to O'Connor's concern about the extent of work involved—an answer that has since been quoted famously in previous accounts of O'Connor's life: 'Railways, Harbours, everything'.[9]

Forrest, the politician, was bargaining with O'Connor. He had up his sleeve a sum of £1,500 for an Engineer-in-Chief, which he had previously estimated in his budget, approved by parliament. Scrupulously honest in his handling of public funds, and in consideration of his own salary being £1,000, Forrest felt it his duty to obtain O'Connor's services at the best possible rate.

But O'Connor weighed the cost of giving up his pension and entitlements and the cost of the move to Western Australia, and wrote his reply on 20 April, which seemed to close the whole episode:

> I conceived from the first that it [the salary] would probably be fixed at more than £1000 p.a.—the words 'not under £1000 p.a.' leading me to that conclusion. Bearing in mind therefore that I should be sacrificing the possibility of a pension, it would certainly not be wise for me to throw up here under present circumstances for less than £1,500 a year.

He then cites the salaries for engineers in other colonies in Australia, and how their salaries were rated according to the number of responsibilities they held. He ends:

> While thus declining the offer which you have made me, I beg at the same time to thank you very much for your kindness in making it, and also to apologise for the trouble to which you have been put through my first cable not being specific.[10]

O'Connor's negotiations with Forrest must have cost him dearly in terms of stress and uncertainty. But this was only one side of the drama unfolding. He was also conducting correspondence with his minister and the Premier in an effort to get his position clarified and his entitlements confirmed. It is this correspondence that shows clearly how short-sighted the government was in dealing with his case, and how bitter O'Connor had become at the treatment given him. As he writes in one document to the Premier:

> To continue to exist in this position of constant anxiety such as I have been doing for the last two years is, I think you will admit, both painful and humiliating, and if the whole wretched struggle for existence is to go on again during the next session [of Parliament], with the almost certain prospect of my losing office, then it would be scarcely bearable.[11]

He explained to the Premier that he had received an offer from Western Australia but that he was reluctant to accept it: 'The prospects offered have no advantages over my present position, in fact they are comparatively disadvantageous, but such as they are I could probably attain them'.[12]

The reply from the Premier, signed by the Secretary to the Cabinet, was terse and as cold and unsympathetic as the sentencing of a hanging judge. Acknowledging receipt of O'Connor's letter, it went on:

> I have to inform you that the Government has considered the matter and has decided to place on the estimates the sum of £750 as salary for your office of Marine Engineer.[13]

O'Connor scribbled on the bottom 'present vote £800', indicating a reduction in salary of £50 in the coming year.

It is not hard to imagine how injured O'Connor believed himself to be on receiving the memorandum. After twenty-five years' loyal service, far from having his position improved his worth had been re-evaluated to save £50. 'I have received exceptionally rigid treatment', he wrote later in what seems to be a remarkably restrained judgment.[14]

The Premier's unsupportive reply was dated Friday, 17 April 1891, and would have been read by O'Connor in his office on the following Monday. Perhaps this was one of the occasions when, as his daughter Kathleen describes, O'Connor was so angry that he jumped on his hat.

Anger at his beggarly treatment and humiliation was clearly justified, and perhaps at this point, like the psalmist, O'Connor cried out to the Lord to deliver him from the depths. If so, his prayer was answered with remarkable expedition.

His deliverance arrived the following day, Tuesday, 21 April, in the form of an unexpected cable from Forrest with a revised offer: 'Would you accept permanent appointment at £1,200 per year under Superannuation Act which entitles you to one-sixth salary after ten years service, and one-sixteenth for each additional year'.[15]

This was O'Connor's life-belt, his longed-for restorative. Without further hesitation, he cabled back his acceptance.

In the ten days remaining to him, he set about obtaining what he saw as his entitlements after twenty-five years' service. In his handwritten notes, which survive, he estimated that 'on a liberal interpretation' he would get, if he retired at that time, £2,333, which would include a year's leave; but if construed on 'a narrower profile basis, allowances would be £1,533'.[16] He confirmed in his submission that he had not had one week's leave of absence in twenty years, and underlined 'one week'. An immediate reply from the Cabinet Secretary, quoting the Premier, again denied him long-service leave. O'Connor wrote to his minister on the same day, restating his case fully. On the following day, once again the Premier's Office refused the leave.

This intransigent attitude on the part of the government arose partly from a legalistic objection to O'Connor's delay in confirming his resignation in writing. But on 29 April he defended himself by saying that he had never made any secret of his position and had discussed it openly with ministers, asking for, and following, their advice. He had postponed his resignation on advice, he said, because he feared that it might create a technical bar to the government showing any consideration in regard to his past service.

Finally, on the same day, 29 April (the correspondence must have been hand-delivered between departments in haste), the Cabinet Secretary issued an authority for O'Connor to receive twelve months' leave of absence on half-pay. Attached was a treasury voucher authorizing payment for half of £800, payable on the last day of the month.

O'Connor's estimate of what was truly owing to him had been ignored.

The *Wellington Evening Post* of the same day carried the news of O'Connor's resignation and his new appointment in Western Australia. There was no doubting where the newspaper's sympathies lay: 'We congratulate him on getting out of the New Zealand Service in its present condition, and securing such a magnificent professional opening as his new appointment affords'.

On the last day of April 1891, O'Connor closed the door of his office for the last time, and ended his pioneer engineering career in New Zealand after twenty-five and a half industrious years, not with the gratuity he had relied upon but with a token £400—well below what might justifiably have been owing to him.

April had indeed been for O'Connor 'the cruellest month'.

PART III
AUSTRALIA, 1891–1902

12

O'CONNOR HAD LITTLE more than a week to attend to family business before leaving New Zealand. Nine days after quitting his office, he bade farewell to his adopted country of twenty-six years from the deck of the SS *Rotamahana* as it steamed out of Wellington Harbour on a southerly course down the east coast of South Island, bound for Melbourne. He was accompanied by three of his children: his eldest daughter, Aileen, Frank and Bridget; Mrs O'Connor remained behind with Kathleen, Eva, Roderick and Murtagh, all of whom would follow later.[1]

The ship called at Lyttelton. During its brief stay, O'Connor probably called on some of his old friends in Christchurch, but would not have visited Avonside. The children's grandparents had both died six years previously, within two months of each other. Coincidentally, George Julius, the eldest son of the new Archbishop of Christchurch, who would later marry O'Connor's daughter Eva in Fremantle, was then studying engineering at Canterbury College. It is unlikely that O'Connor would have known the Julius family at the time.

The *Rotamahana* rounded the Bluff—the southernmost point of South Island—at 5.30 p.m. on 15 May 1891 and headed westward. It 'experienced strong westerlies with a high confused sea'.[2] What thoughts O'Connor had as he stood on deck in the mildly rough weather, watching the country he had adopted slipping over the horizon, we can only guess. In a letter to a friend in Christchurch, he later denied that he ever lost sight of New Zealand:

> although circumstances forced me to leave there, and feeling, perhaps wrongly, that I did not quite get justice there, I still feel, and I believe I will always feel, the warmest interest in her welfare generally as well as in the welfare of many friends from whom it was a great wrench to part.[3]

The children would have felt a greater wrench at leaving. Kathleen tells us, years later, that she did not want to settle in a new country but her father 'was so absorbed in his work that he hardly seemed to realise the change of country'⁴—a judgment hard to equate with his own statement quoted above. However, there being no work to absorb him on the ship, he would have experienced his first holiday in twenty-five years. Kathleen admitted that he often felt the need of his old friends and 'suffered a loneliness indescribable'.⁵

The ship entered Port Phillip Bay late in the evening of 19 May and came alongside Queens Wharf early the next morning. The Victorian Parliamentary Railway Committee was then conducting an inquiry into the colony's railway system and O'Connor had been invited to give advice—probably privately, because he was not called as a witness.

The O'Connors transferred to the P&O ship *Massilia*, which was already in port when their New Zealand ship arrived. The new Engineer-in-Chief cabled the Western Australian Government that he would arrive in Albany about 1 June. The news was duly noted in the *West Australian* newspaper of 21 May. This important, highly paid Forrest appointee from whom so much was expected was awaited with curiosity and eagerness—and doubtless with a touch of scepticism in some quarters.

None of the family would have had an accurate picture of what was in store for them in Western Australia, a colony so remote from the New Zealand they knew, and so anonymous. Perhaps they based their expectations on their received knowledge of Melbourne and Sydney, which were bustling Victorian cities much in advance of the West at that time. It was on those cities that the popular image and wealth of Australia rested; the West was comparatively unknown and rated unimportant beyond its own boundaries. Western Australia had already been dubbed the 'Cinderella State', an epithet that would taunt it for the next hundred years.⁶ But if that description was thought to be justified even by the middle of the twentieth century, in the early 1880s a more accurate comparison might have been with one of the ugly sisters. The newest independent colony had only recently emerged from years of stagnation, underdevelopment and diminishing population, rescued—albeit reluctantly—by a short period of convict transportation and early gold discoveries. Although transportation had ended twelve years before the O'Connors' arrival, and the economic tide had begun to turn under the leadership of John Forrest and with the introduction of independent constitutional government, the task of pulling the colony out of the doldrums was still in the future, and immense. Cinderella's golden coach was approaching from the direction of the Murchison, the Kimberley and most recently the Yilgarn, but would not reveal its richest treasures until the glory days of 1893 onwards. As O'Connor's fellow Meathman the poet and convict John Boyle O'Reilly so presciently described Western Australia in his poem written in the bush twenty years before:

> Virgin fair thou art,
> All sweetly fruitful waiting with soft pain
> The spouse that comes to wake thy sleeping heart.[7]

The momentary temptation to see O'Connor as the imagined spouse in O'Reilly's poem should be resisted; more likely the instrument for awakening the sleeping heart was the serendipitous coming together of Forrest and O'Connor at the same time as new wealth from the major gold discoveries of the decade.

Some of the factors attendant upon O'Connor's arrival in Western Australia were curiously similar to those existing in New Zealand when he arrived there in 1865. Both colonies were on the eve of unprecedented economic development arising from gold discoveries which, in turn, necessitated an expansion of public works financed by overseas loans. New Zealand's transformation to centralized government happened slowly after O'Connor's arrival there; Western Australia's first independent parliament, under its new Constitution, was six months old when he arrived. John Forrest, who had been returned unopposed in his Bunbury constituency on a strong platform of expanding public works, was appointed Premier by the Governor, Sir William Robinson. When the first parliament met on 20 January 1891, Forrest's ministry, now remembered mainly in street names and the suburbs of Perth named after them, included William Edward Marmion, Commissioner of Crown Lands; Septimus Burt, Attorney-General; George Shenton, Colonial Secretary; and Harry Whittall Venn, the first Commissioner of Railways and Public Works, who was to prove a loyal ally of O'Connor's. These ministers, whose wealth derived from their business interests and large landholdings, formed a ruling circle and wielded considerable influence in the colony. They were for the most part cautious, shrewd and conservative. They dined together at the elite Weld Club and met at vice-regal social functions. There was no formal opposition party until 1894, and for the first four years it was to prove ineffectual in the face of the dominant leadership of Forrest.

John Forrest, like Julius Vogel in New Zealand, persuaded his more cautious parliamentary colleagues that the only way forward was to finance public works by raising huge loans on the London money market. The new Constitution empowered the colony to do this, rather than having to rely on private capital raised by land-grant or other syndicates. Forrest's first Loan Bill to raise £1,336,000—an unprecedented figure in those days—was passed in February. The stage was now set for the arrival of Forrest's chosen one who would 'change the order of things'.

The *Massilia* berthed at Albany late on a stormy Saturday evening, 30 May, necessitating the O'Connors' staying overnight in Albany. The hotels were overcrowded and uncomfortable owing to a flood of hopeful diggers arriving in Albany for the goldfields; however, in all probability, the O'Connors would have been rescued by the Wright family, who lived in a double-storey house on the Esplanade overlooking the harbour. J. Arthur Wright had been Director of Public Works until his resignation in 1889. He and his family were to play hosts to the second contingent of O'Connors when they arrived in Albany two months later.

Albany, which boasted the original landing place of European settlement in the colony, three years before Perth, was then a straggling township of some 3,000 residents, 250 miles south of the capital and more than 340 miles distant by rail. Its importance centred on its superb deep-water sheltered harbour, which was the port most frequented by international shipping in preference to Fremantle—a widely acknowledged impediment to economic development and easy communication.

Renowned for its scenic beauty in good weather, Albany must have presented to the O'Connors a dreary, cold and windswept welcome in its winter mood. At that time of the year, successive rain-bearing depressions sweep in from the southern ocean—an uncomfortable reminder of the stormy West Coast of New Zealand. The summer beauties of King George Sound, with its inner land-locked harbour, would have been hidden from the visitors by thick weather. At night the noise of the frogs, which so startled Trollope that he thought 'the hills were infested with legions of lions, tigers and bears',[8] would have been muffled for the O'Connors by the torrential rain on the roof.

Sunday, like the previous day, dawned cold and wet. The rain continued all day and night. Streets were flooded, and considerable damage was reported.[9] The Perth train, incongruously described 'express', left at 6.00 p.m. on the Sunday evening, arriving in Perth at 1.50 p.m. the next day—a journey of more than twenty hours. The O'Connors' fellow passengers on the *Massilia* had included prospectors from the eastern colonies bound for the new gold finds in the Yilgarn. Albert Calvert, explorer, author and prospector, described the crowds at Albany railway station, rushing about in search of luggage, scrabbling for a window seat in a stuffy train, passengers sitting on each other's laps 'like trussed fowls' and the train making its 'unlovely journey to Perth' via the isolated country towns of Katanning and Beverley.[10] Although the first part of O'Connor's journey—as far as Beverley—was on one of the privately owned railways, it was surely ironic that the new general manager in charge of government-owned railways should have experienced such a graphic introduction to the confused and disjointed system that confronted him—and have ample time to turn over in his mind possible strategies for improvement.

After changing trains at Beverley onto the government line, the express puffed its way into Perth Station in the afternoon of Monday, 1 June, the Foundation Day holiday weekend. Among the celebrations staged were a regatta at Fremantle, and the first race meeting of the Fremantle Turf Club near Woodman Point, 5 miles south of the city. On that occasion, one thousand return tickets were issued from Perth to Fremantle for the events, an increased number travelling by train because the steamers, which would have ferried passengers to the race meeting, had been cancelled owing to recent storms having destroyed the jetty close to the race track. Later a Grand Ball was held at the Perth Town Hall, unhindered by the largely good-natured workers' demonstration and procession through the city in support of the eight-hour working day—a movement O'Connor would strongly support for his workers at a later date.

Coincidental with O'Connor's arrival that weekend, there was an additional reason for celebration on the part of the Premier, his supporters and—if the *West Australian* report is to be believed—the majority of citizens of Western Australia. The Governor had announced that Her Majesty the Queen was pleased to confer the title of Knight Commander of the Order of St Michael and St George on John Forrest. Henceforth he would be known as Sir John. Only in the last few months of his life in 1918, while a sick man, would he be elevated to the peerage and become known as Lord Forrest of Bunbury.[11]

Perth in 1891 was a flat, sprawling town characterized by wide streets, large empty spaces, and one-storey verandahed shops whose ornate facades belied the utilitarian structures at their rear. Only a few main streets had been macadamized; others were dusty from compacted sand and deeply rutted by carriage wheels. There were a few substantial buildings, chief among them the Town Hall, the new government offices on St Georges Terrace, the Anglican cathedral and Government House. But the general impression, heightened by the number of horse-drawn carts and cabs, and the recently erected telegraph poles and wires, was of an unloved frontier town. One traveller commented that 'You feel yourself more out of the world in Perth than in Siberia'.[12] And even the parochial *West Australian* newspaper admitted in an editorial that 'The dreariness of Perth as compared to other capital cities of Australia has frequently formed the subject of comment'—but then, by way of compensation, it added that established residents fortunate enough to have a wide circle of friends and a social round found 'life a good deal more than tolerable'.[13] The physical appearance of Perth would change substantially in the decade ahead, but C. Y. O'Connor and his three travel-weary children must have viewed the scene through the train window with despair, comparing it unfavourably with bustling, scenic Wellington, where the rest of the family had so recently been left behind.

The O'Connors probably did not alight at Perth on this occasion; the train terminated at Fremantle and stopped en route at Claremont, where lived their host for the first days, Colonial Architect George Temple-Poole.[14]

An early indication of O'Connor's devotion to his work was the speed with which he took up his duties after a wearisome journey. On his arrival, his short break from work was at an end. Leaving his children and his domestic arrangements in the care of Aileen, then aged 17, O'Connor—as reported by the *West Australian*—started work the very next day, 2 June. The paper astutely listed his duties, describing them as 'of a multifarious and pressing character', and concluded knowingly, but without exaggeration, 'It is clear that Mr O'Connor will have his hands fully occupied for some months to come'.[15]

13

OVER THE HUNDRED or more years that have elapsed since O'Connor first began to work for the Government of Western Australia, business and industry have become increasingly compartmentalized, with each functionary an accredited specialist working within a well-defined area of operation. This accepted modern practice makes all the more remarkable and seemingly burdensome the 'multifarious duties' that faced O'Connor when he first entered the Public Works offices, housed at that time on the first floor of a new building on the corner of St Georges Terrace and Barrack Street. Completed a year before O'Connor arrived, at a total cost of just over £31,000, it was—and still remains—one of the most handsome and most impressive buildings in Perth. The classical-renaissance four-storey frontage on St Georges Terrace housed, at that time, the general post office on the ground floor. O'Connor would have entered the government offices through the side entrance on Barrack Street and mounted the main jarrah stairway.[1] Any elation he may have experienced at the grandeur of his new surroundings may have given way to disappointment when he discovered that he had practically no trained staff, other than George Temple-Poole, a trained architect and engineer who had served as Director of Public Works in the old colonial administration. He had been appointed Assistant Engineer-in-Chief until O'Connor arrived.

O'Connor also found on his desk a dauntingly overloaded in-tray.

As both Acting General Manager of Railways and the Engineer-in-Chief, O'Connor was responsible to the minister for supervising work in progress, proposing and planning new routes, designing bridges, railway buildings and crossings, maintaining harbour jetties connecting with railways, and ensuring there were adequate supplies of water and fuel for rolling-stock. He was also expected to tackle the loss the railways of the colony had consistently made and, by careful management and by identifying weaknesses in the system, turn railway operations into a profitable enterprise. In all of these duties, O'Connor's practical experience in New Zealand and his later executive role as Under Secretary in Wellington responsible for all the colony's railways gave him a unique capability and an authority that quickly earned for him the respect of his minister, Harry Venn. The combination of his engineering experience and economic management, and his intuitive grasp of detail, would surely be the envy of any modern-day Harvard MBA.

But before considering in more detail O'Connor's railway work, let us glance at some of the other equally pressing files in his in-tray on that morning when he stepped into his office after walking—or more likely riding in a hansom[2]—up the hill from Perth's railway station.[3]

Equally if not more urgent than the reorganization of the government railways was the long-debated question of building a suitable port for Fremantle capable of servicing world shipping and providing easy and efficient rail communication with the hinterland—and particularly with the fast developing Eastern Goldfields. A rock bar and shallows that dried out at low water blocked the entrance to the Swan River, and those freight ships that risked the open and often dangerous conditions outside berthed at the 1,006-yard jetty, known for obvious reasons as the Long Jetty.[4] Conditions prevailing there were notoriously uncomfortable and the subject of numerous complaints. The most oft-quoted was that of the American sea-captain D. B. Shaw of the sailing ship *Saranac*, who wrote:

> It is a terrible place. No place to put a vessel. No shelter whatever. It is certainly the worst place I or anyone else ever saw. And any man who would come or send a ship a second time is a damned ass.[5]

Less blunt, but no less critical, was the Chief Harbour Master, who wrote in his annual report a year before O'Connor arrived:

> Considerable dissatisfaction had been expressed lately in London at the delay in discharging steamers at Fremantle...it has been stated that freight [charges] would have to be considerably increased before steamers would be allowed to load for this port. What shippers require is such shelter as will enable their steamers to discharge a certain quantity of cargo in a given time without detention on account of delays caused by bad weather.[6]

In bad weather, ships had to lie off in Gage Roads and discharge cargo—weather permitting—into lighters. The lighters would then off-load at the Long Jetty, the freight conveyed by train, or by cart across Fremantle to the Town Jetty beyond the bar in the river—a lengthy and expensive operation. Even in good weather, the Long Jetty could only accommodate vessels with a draft less than 12 feet. Mail steamers avoided Fremantle altogether, preferring distant Albany, which, as noted previously, seriously inhibited trade and communication. That Fremantle required a modern, safe harbour was acknowledged by all. But it was the siting of that harbour, and the cost, that were hotly debated. O'Connor's first intimation of the strong feelings in the community on the matter would have been his likely meeting with a fellow first-class passenger from Adelaide on the *Massilia*, Mr J. S. Yuill. Yuill was the Australian manager of the Orient Steamship Company on a visit to Western Australia to inspect the harbour, meet a business delegation, and urge the government to improve accommodation to an extent that would allow his company's ships to berth there. And again, on the morning of O'Connor's arrival in his office—and therefore perhaps deliberately timed—the leader in the *West Australian* was urging the government to address the problem:

> To suppose that the metropolis, and with it the most thickly populated portion of the colony, will be content always to owe their contact with other countries to a route which cannot be described by any milder epithet than that of circuitous, is to assume that they have no ambition to advance beyond their present position.[7]

As early as 1875, the then colonial administration had sought advice from the eminent British harbour engineer, Sir John Coode. Without visiting the colony, Coode made recommendations based on Admiralty Marine Surveys, although neither of his two suggestions were adopted at the time.

Ten years later, he was again consulted and on this occasion visited the colony on his way back to England from New South Wales. He spent five weeks in the July and August of 1885, 'making inspections...examining Gage Roads, Owen's Anchorage and Cockburn Sound and investigating the physical characteristics of the Swan River between Perth and the sea'.[8] On leaving, he requested that further work be carried out in his absence to determine the composition of the Parmelia and Success banks. Back in London, Sir John wrote his recommendations based on his own observations and the data collected after he had departed. He rejected outright three suggestions that had been strongly mooted: the cutting of a channel from what is now Leighton Beach across into

Rocky Bay, the point where the Swan sweeps around closest to the sea; the provision of a harbour in relatively sheltered Cockburn Sound; and siting of the harbour in the Swan River itself.

With regard to the Rocky Bay proposal, Coode believed that a sandbar would develop at the ocean entrance to the channel and that maintaining a suitable and safe approach for shipping would be too costly. With the second proposal, he was also worried about the southerly movement of sand along the coast and the need for continual dredging. On the matter of a town harbour in the Swan River, he stated that this proposal had 'engaged my anxious and careful consideration'. But he concluded that 'Rock removal would be a work of considerable difficulty and attended with an expenditure which would be altogether disproportionate to the benefits derived from the deepening'. He then went on to say that the same problems arising from sand travel in the first two cases would be so adverse that it would be 'quite impractical to treat the existing entrance to the Swan with a view to the formation and maintenance of a deep-water approach with any degree of success'.[9]

Coode's own recommendation, as a first stage in an ongoing plan, was to build a solid masonry wharf curving out in a south-westerly direction from Arthur Head. The estimated cost was £448,000, with an additional £47,000 for dredging to a depth of between 22 and 29 feet at low water. The second stage was more ambitious, calling for further dredging to a depth of 34 feet and increased wharfage. The matter was further complicated by another popular scheme, supported by Premier John Forrest, to build the harbour at Owen Anchorage, a sheltered bay south of Arthur Head and just north of Woodman Point. Although Sir John Coode's consultancy fees had already amounted to £3,477, the Premier again insisted on seeking his advice, this time on the Owen Anchorage scheme. But before Sir John's response could be received, the new Engineer-in-Chief was appointed. H. W. Venn, the Minister for Railways and Public Works, very properly argued that with O'Connor's imminent arrival, no decision should be taken before the new appointee had had time to consider all the various proposals.

Thus O'Connor was confronted with insistent demands from two quarters: on the one hand, the complex matter of the best and most affordable port for Fremantle; and on the other, the depressed and inadequate condition of the railways, which had been made worse by a recent spate of accidents directly attributed to steep gradients, unsuitable rolling-stock and light engines attempting to pull too heavy loads.[10] Scarcely had he settled into his office when he was required to attend a meeting at which his minister received a deputation of citizens complaining about the siting and dangerous inadequacy of the city's railway crossings. And in between times there were insistent demands from regional centres pleading for the upgrading of their port facilities.

O'Connor's domestic arrangements, in the absence of Mrs O'Connor, were left in the hands of Aileen. She was evidently competent beyond her years, and was much admired. Her sister Kathleen says she 'was worshipped and she seemed to be able to do anything'.[11] For the first year in Fremantle, the family leased 'Park Bungalow', 7 Quarry Street, then owned by the Colonial Surgeon, Dr H. C. Barnett. When Dr Barnett returned to occupy his house in January 1893, the O'Connors rented, for a few months only, 'Yeldam House', Lot 300 in Cantonment Street. Afterwards they leased 'Plympton House', a short distance away in Beach Street, overlooking the Swan River (now the harbour). The family would move back into 'Park Bungalow' in 1896 for four years, then return to 'Plympton House' in 1900.[12] It is curious that O'Connor, who received the top civil service salary, did not—as so many of his colleagues, politicians and leading citizens did—buy property in Western Australia but was content with moving back and forth to rented property. Was he reluctant to put down roots? And did he intend moving back to New Zealand—or Ireland? We cannot know.

'Park Bungalow', dating from the early 1870s, was built of local limestone on raised ground overlooking Fremantle Park and within sight of the present-day Fremantle Arts Centre (then the asylum). It had high ceilings, a library, dining room, music room and drawing room, bedrooms and a wine cellar. A housekeeper's quarters were located on a lower level. What would have been an important consideration for O'Connor was the provision of stabling for horses, a feature also of 'Plympton House'. Both 'Park Bungalow' and 'Plympton House' would have been substantial residences for those times, fitting homes for someone in O'Connor's position.[13]

Horses were always O'Connor's means of relaxation and escape:

> To throw off all care was to be riding a favorite horse, accompanied by one of his daughters—a blissful hour in the morning between 7.30 and 8.30, then a bath and breakfast, catching the train to Perth for a long day's work.[14]

O'Connor employed Irish immigrant boys from farms in Ireland to look after his horses:

> They were part and parcel of the family and the devotion of these boys was the only word to describe what they were willing to

undertake for us; from sailing us in the mornings to keeping the horses groomed and ready at any hour, Saturdays and Sundays included. They finally all became policemen, some very advanced in the force.[15]

Among O'Connor's papers is a list of meets of the Fremantle Hunt Club, which were generally held on Saturdays during his first year in the colony, and a letter showing that he had an interest in rowing, being re-elected unanimously as vice-president of the Fremantle Rowing Club. Kathleen O'Connor wrote that her father 'never accepted invitations, but liked people to come to the house',[16] and a close friend said that O'Connor 'did not cultivate the men his contemporaries might think it was useful to know'.[17] This may be interpreted to mean that O'Connor kept much to himself and did not mix socially. This is not strictly true, as the records show. As a senior public servant, close to the Premier, he attended garden parties and receptions at Government House and more frequently at the Forrests' hospitable home in St Georges Terrace. Within days of his arrival, he was elected a member of the exclusive Weld Club, whose premises on Barrack Street across the road from his offices were frequented by the leading landowners, legislators and legal fraternity. His sponsors were his minister, Harry Venn, and Sir Edward Stone, the stately, austere Chief Justice. Perth society was only too anxious to examine the Premier's new Engineer-in-Chief, known to have aristocratic connections, and although the old patrician families were generally suspicious of newcomers, O'Connor's unfailing courtesy, impeccable manners and consideration of others led to close and lasting friendships. Within a month of his arrival, he was attending a 'Grand Mayoral Ball' at the Town Hall at which Sir John Forrest and most of his Cabinet were present. O'Connor partnered Mrs Burt in the opening quadrille, her husband, the Attorney-General, being overseas and Mrs O'Connor not yet having arrived from New Zealand. 'Of the supper it may be said that no better has been served in Western Australia.'[18]

The weather that July 'prevailed with unusual severity', and rainfall was heavier than had been experienced for many seasons. The city was covered with water and mud. The *West Australian* of 7 August complained of the state of the city's roads: only a small portion of St Georges Terrace was yet metalled, and most other places were covered with mud or a verdant carpet of short-lived grass.

Those first months of O'Connor's employ were for him a time of familiarization with the scale of his responsibilities. In the first week in July, he made his first visit to Bunbury to inspect the route for the new South-Western

Railway, which had been surveyed before his arrival, and expressed himself well satisfied with it. On 13 August, he left for Geraldton by sea in company with his minister, Venn, to determine the extent of public works proposed for the Champion Bay district. This trip would have coincided with the arrival of Mrs O'Connor and the children on the SS *Omruz* at Albany in a winter storm on 15 August. For O'Connor, his work and loyalty to the government took precedence over family matters. But this should not be interpreted as a lack of affection or interest in his home life. Wrote his daughter Kathleen:

> there are hundreds of stories to show that we as a family were always encouraged to say something bright and to the point. If the answer happened to be really clever we received half a crown. From all this it can be realised that C. Y. O'Connor had much pleasure in his home in spite of the fact that we were all rather harem-scarums.[19]

Mrs O'Connor, Eva, Kathleen, Roderick and baby Murtagh remained in Albany with the Wright family for a number of weeks. Just how long is not certain, but we know they were in residence in Fremantle by October 1891 because Susan O'Connor accompanied her husband to a garden party given by Lady Forrest on the 28th. Aileen was also present and there she met the Governor's secretary, C. Y. Simpson, who later became her husband.

14

WHENEVER THE NAME of C. Y. O'Connor is mentioned, it is invariably associated with the Goldfields Pipeline bringing water from the coastal range to the arid Kalgoorlie region—and then, by way of an afterthought, his responsibility for the design and construction of Fremantle Harbour may be remembered. As dominant as these achievements were, seldom is acknowledged his work in reorganizing and planning extensions and improvements to Western Australia's railway system, work that is neither understood nor appreciated. O'Connor's experience and grasp of both the practical problems of pioneer railway construction and wise economic management were unsurpassed at that time. Together with Venn, Commissioner for Railways and Public Works, a strong ally and defender of O'Connor's proposals in parliament, O'Connor turned the colony's railway system from loss and inefficiency into expansion and commercial success. But it did not all happen overnight, nor without opposition and some-times rancour.

When O'Connor arrived in the colony, there existed 188 miles of government lines. The so-called Eastern line linked Fremantle with Perth and the main agricultural areas of the Swan Valley. The Bunbury line had been surveyed and construction started. The line to Albany via Beverley was then owned by the Western Australian Land Company, and a similar privately owned line was being constructed to link Perth with Geraldton. O'Connor saw that in order to provide a better service and greater safety and profit, the Eastern line had to be re-routed and upgraded, and new rolling-stock purchased.

Before O'Connor arrived in the colony, Forrest had argued for rapid and cheap means of communication: a farmers railway from Perth to the South West (Forrest's own constituency); the extension of the Avon Valley line to the Yilgarn goldfields; and a line from Geraldton to Mullewa. He saw it as the government's

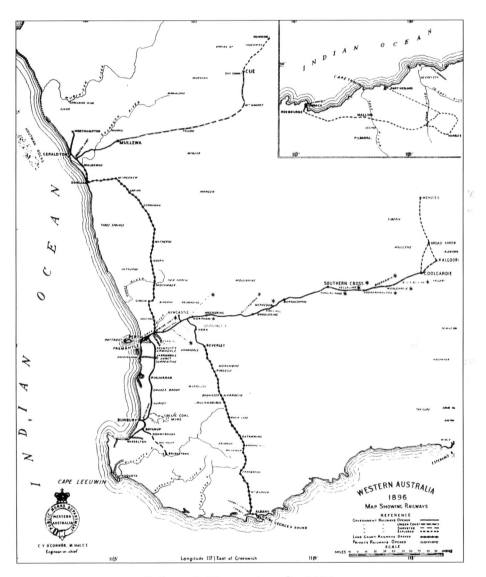

Railways in Western Australia, 1896.

Original held at SROWA

duty 'not to sit and wait for traffic, but to create traffic'. The policy of the new government, he claimed, would be to open up the country and facilitate development.[1] O'Connor's approach was the opposite: he argued that railways were only justified when it was anticipated that revenue would cover expenditure, including the interest on loan repayments. Unlike politicians, he was not subject to pressure from wealthy constituents nor motivated by sectional interests, but mindful of the good of the colony as a whole. There would be other occasions when O'Connor's advice ran contrary to Forrest's declared views, but it is a tribute to Forrest's responsiveness that he was persuaded by O'Connor's arguments and strongly supported him. Forrest was sometimes 'slow to think in new ways, but having decided on a course of action, especially if it looked bold…he moved promptly'.[2] O'Connor, he stated generously, was 'a most excellent officer, in every way reliable, and a man of great ability'.[3] This view was not shared by Forrest's brother Alexander, member for the West Kimberley constituency and engaged in numerous mining and commercial interests throughout Western Australia. His was the first of many hostile voices to be raised in opposition to O'Connor's power, his success and his tendency to disregard civil service red tape:

> We know that the Engineer-in-Chief represents and rules not only the Works and Railways, but the whole colony. He not only rules my Hon. friend's [Venn's] department, but every other department of the public service; and as I stated the other day, in ten years time we may as well hand over the whole colony to this gentleman from New Zealand who will no doubt show us how to spend our money.[4]

Thus, within a short time of O'Connor's arrival in the colony, he began to ruffle the feathers of the entrenched ascendancy, particularly those whose business interests, whether in commerce or industry, were thought to be threatened. As scrupulously honest as he was, some, who were accustomed to preferential treatment for members of the 'old-boy network', felt they were disadvantaged by his decisions concerning the letting and settlement of government contracts, the expenditure of funds and the influence he wielded. There was criticism also of his methods. Indeed, there is some justification for believing that O'Connor, in his professional life austere, uncompromising and never one to suffer fools gladly, brought some of the criticism on himself. For example, when tenders for an extension of the Eastern line—the Southern Cross Railway contract—closed on 11 June 1895, the fifty hopeful tenderers assembled in O'Connor's office to learn who had been successful in gaining the contract. O'Connor 'stood at his chair with the tender box on the table in front of him'.⁵ Next to him was the minister, Harry Venn. All was silence as the box was opened; each movement of O'Connor's was watched with close attention. He slowly withdrew the envelopes one by one, opening them, scanning the contents intently and noting down the figures. Then, taking up one, he leaned towards Venn and said in a low voice:

> I do not think we need to proceed further. For the Southern Cross
> Railway Contract seven tenders have been received, the lowest tender
> is by John and Adam Wilkie for £64,125.13s.7d.⁶

This was so far below the average tender figure—under half the highest of those presented so far—that suspicions were raised. The unsuccessful tenderers (the Wilkies were not present) then rose from their seats and filed out of the office, struck dumb at first, but later there arose a babble of disgruntled voices.⁷

The Wilkie brothers had come from New Zealand, and had previously tendered successfully for one of O'Connor's railway constructions there. Here, it was supposed, was yet another example of O'Connor favouring his New Zealand friends.

On his arrival in Western Australia, O'Connor had argued successfully for an increase in trained staff and had recruited three skilled and trusted officers from his old country: F. W. Martin, appointed engineer-in-charge of existing lines; W. W. Dartnell, maintenance engineer; and A. W. Dillon Bell, inspecting engineer. The choice of staff was in O'Connor's hands, but the numbers employed had to be approved by parliament, and at a later date, O'Connor's decision led to an outburst of criticism from the Member for Perth, T. G. Molloy:

> The Engineer-in-Chief seems to have complete control. He seems to
> have the management of the affair as though it was his own company,
> that he was not a public servant; that he seems to be omnipotent and
> employ whom he thinks fit. In many instances, it is thought, he does
> so in a reckless way, and with extravagance in the carrying out of his
> duties. He has particularly favoured one colony, the colony from
> which he has come…undue preference has been given to persons
> hailing from New Zealand.⁸

O'Connor's minister, H. W. Venn, led a stout defence of his Chief Engineer, replying that in view of the lack of trained personnel in Western

Australia it was only reasonable to suppose that O'Connor would recommend men in whom he had most confidence:

> If he came from Victoria, he would recommend the appointment of Victorians as being men whom he knew and could trust. If the Government had known of men of equal ability whom they could appoint, they would have been appointed. But they did not.[9]

Venn went on to say that O'Connor was under the greatest possible strain in having at times to answer and deal with important questions put before him by resident engineers in different parts of the colony, and it had become impossible for him to attend to all the many duties in his office and at the same time travel all over the colony to inspect work in progress.

In spite of these bursts of criticism, inspired for the most part by jealousy and suspicion of outsiders, there was much support for O'Connor, who was recognized as an effective and valuable asset to the colony. John Forrest declared in parliament that the colony was fortunate to have him. Even Alexander Forrest, who on one occasion, under parliamentary privilege, mendaciously claimed that O'Connor had been dismissed from government service in New Zealand, admitted grudgingly that the railway service had been greatly improved under O'Connor's control.[10]

The *West Australian*, ever quick to support the government, praised O'Connor and in one early editorial sympathized with the new Engineer-in-Chief:

> Mr O'Connor's energies should be left entirely free for the scientific designing and supervision of the many important engineering undertakings upon which the colony is entering, and this obviously cannot be unless he has the assistance of an experienced and thoroughly qualified business manager—an officer in whom he can have full confidence.[11]

O'Connor's early recommendations included improving the gradients over the Darling escarpment on the Eastern line; substituting heavier, 60-pound rails for the existing 46-pound rails to the yard, and purchasing heavier locomotives; and upgrading and moving the railway workshops from Fremantle.

The relocation of the workshops would take ten years of bitter wrangling and persuasion, and it was nearly two years after O'Connor's appointment before Forrest confronted the Legislative Assembly, requesting £111,743 from loan funds towards the £133,000 estimated by O'Connor for upgrading the lines:

> For a long time I tried to withstand the appeals made by the Commissioner of Railways [Venn], in fact I put him off a good many times…But it has been brought home to the Government that we should do something to improve the grades on the Eastern Railway. For some months past this has been forcibly brought before the Government by the Engineer-in-Chief.[12]

The work on upgrading the lines was commenced in October 1893, and the Mahogany Creek Deviation contract was let in February 1894. This latter work necessitated a succession of deep cuttings through solid granite rock, the construction of six bridges, and the boring of a tunnel at one section under the Darling Range for a length of 1,089 feet—the only railway tunnel in Western Australia until very recently. It is this tunnel that is depicted at the base of the O'Connor statue in Fremantle Harbour (see page 2).

O'Connor also introduced a new accounting system showing revenue and expenditure on each line. His criticism of what he believed was the outdated system then in place led to a conflict with the chief clerk in the Auditor-General's office. A Select Committee of Inquiry was appointed in 1896 to arbitrate, and severely reprimanded O'Connor, although much of his other work was commended.[13]

O'Connor's first report on the working of the railways for 1891 indicated that for the expenditure of £133,000 on improvements, a saving of £19,000 would result. When the final figures were available, they showed a saving much in excess of O'Connor's estimated figure.[14] O'Connor's subsequent reports for the years 1892–95 showed that goods were being handled more efficiently and the proportion of revenue to expenditure had increased significantly.[15]

In March 1892, the *West Australian* reported that it was in receipt of the drawings and specifications for the Bunbury section of the Perth–Bunbury Railway, prepared under the direction of Mr C. Y. O'Connor. The drawings, numbering ninety-nine sheets, were 'deserving of the highest praise and the information they afford is probably the most complete given by any plans of public works prepared in this colony'.[16]

And while all this work on railway improvements was being advanced, O'Connor was devoting at least an equal amount of time to the question of a suitable port for the colony.

15

MANY DECISIONS OF a political or social nature in history that prove to have been overwhelmingly the right ones in retrospect, and have since brought benefits to succeeding generations, were often decided upon at the time after a closely fought debate when a less satisfactory, much favoured alternative might just as easily have won the day. And in such cases it has been, generally, the genius or determination of one person that has ultimately swayed opinion. A modern example in Australia is the Sydney Opera House, whose existence on its harbour site owes much to the vision and persistence of Sir Eugene Goosens.

It may not be generally realized that the siting of Fremantle Harbour at the mouth of the Swan River is an earlier example. In retrospect so obviously the ideal and economically successful location, it was not thought to be so by the majority and by the experts at the time, and the present harbour would certainly not have existed had it not been for the vision of C. Y. O'Connor. Without his persistence and logical argument swaying adverse opinion, Fremantle would have had a much inferior, less serviceable harbour at Owen Anchorage, requiring continual modification and expensive maintenance to accommodate even the most modest shipping requirements.

On Wednesday, 28 October 1891, Mr and Mrs O'Connor together with their eldest daughter, Aileen, were present at Lady Forrest's garden party, after which they moved across St Georges Terrace to Government House for an evening reception in honour of the officers of the visiting naval ship *Katoomba* (then in all probability at uncomfortable anchorage in Gage Roads). Sir John Forrest was also present at both events, as was O'Connor's minister, Harry Venn. As they were in the thick of exchanging memoranda on the subject of the location of Fremantle Harbour at that time, it is more than likely that they would have discussed the harbour while in the refreshment tent—the refreshments 'served on the most extended scale'.[1]

The Premier's preferred option was for a harbour at Owen Anchorage, 2 miles outside and south of the river entrance. In this he was greatly influenced by Sir John Coode's advice from London, which he had solicited just prior to O'Connor's arrival. Working entirely from charts and data sent to him, Coode advised that suitable channels could be cut through both the Parmelia Bank, so called because the first immigrant ship of that name had foundered there in June 1829, and the Success Bank, named after the naval ship that was also, more seriously, grounded there in November of the same year. It was hardly to be reckoned an auspicious beginning for Western Australia's major port facility. Undeterred by history, even if he had been aware of it, Coode had advised Forrest that

> A channel through the Parmelia Bank, by reason of its sheltered position, would probably be found sufficient, if formed with a bottom width of 250 feet and a depth of 33 feet below summer low-water level. The length of this channel would be about one sea mile, whereas that through the Success Bank would be nearly two miles in length.[2]

Coode then goes on to advise the type and size of dredgers that would be necessary for the work and the length of time that the work would take.

Forrest, obviously encouraged by the support of the world's then most eminent harbour expert, asked his new Engineer-in-Chief to prepare a report setting out the costs of the Owen Anchorage option preparatory to presenting the plan to parliament. O'Connor dutifully complied, concluding that an outer harbour might be built there for £150,000, with an additional £50,000 necessary for dredgers, a railway linking the harbour with Fremantle, and other facilities. But he must have entertained serious doubts about the wisdom of siting the harbour in that treacherous locality. (Any modern-day yachtsman or woman who warily sails between those banks in Cockburn Sound would well understand such concerns.)

O'Connor's growing conviction, contrary to Sir John Coode's advice and after close study of the data, was in favour of a harbour within the river, either by cutting a channel through to Rocky Bay or by dredging the river's mouth. He countered Coode's chief objection to siting the harbour within the river—the problem of sand travel (that is, the continuous movement of sand in one direction from a limitless source)—by suggesting that Sir John had been given inaccurate and insufficient information on which to base his conclusions.[3] O'Connor gained the support of his minister but neither of them could shift Forrest from his favoured option. Wrote Forrest in his own hand:

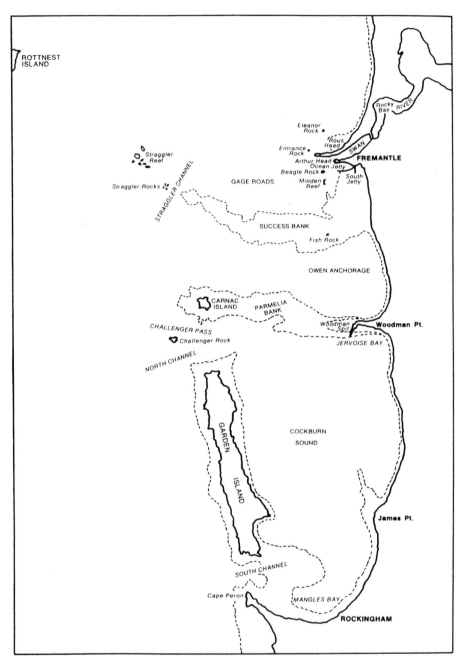

ROTTNEST ISLAND

Straggler Reef

Straggler Rocks

STRAGGLER CHANNEL

GAGE ROADS

Eleanor Rock

Rous Head

Entrance Rock

SWAN

Rocky Bay

RIVER

Arthur Head

Ocean Jetty

Beagle Rock

FREMANTLE

Minden Reef

South Jetty

SUCCESS BANK

Fish Rock

OWEN ANCHORAGE

CARNAC ISLAND

PARMELIA BANK

Woodman Spit

Woodman Pt.

CHALLENGER PASS

Challenger Rock

JERVOISE BAY

NORTH CHANNEL

GARDEN ISLAND

COCKBURN SOUND

James Pt.

SOUTH CHANNEL

Cape Peron

MANGLES BAY

ROCKINGHAM

The seaward approaches to Fremantle, showing Sir John Forrest's preferred site for the new harbour, Owen Anchorage, in relation to the Swan River and the dangerous Success and Parmelia banks.

Map by Geoff Ward (1978)

The harbour entrance and the Rocky Bay Scheme would cost too much and take too long to construct. I think the Government must take its stand upon Owen's Anchorage. I believe we shall have the support of Parliament.[4]

Forrest also had the support of the influential *West Australian* newspaper, then edited by Winthrop Hackett, a colleague of Forrest's in the Legislative Council. An editorial on 25 September 1891 opined:

The proposals for opening up the river belong, in all probability, to a later period of our history. They commit the colony to too high an expenditure. The scheme which has just been sanctioned by Sir John Coode enables us, at what no person can describe as an excessive cost, to bring the largest ships to Fremantle and to bring them there at an early date.

But Venn never wavered in his support of O'Connor's proposals and warned the intransigent Forrest that 'the advantages [of an inner harbour] over every other scheme are *so apparent* and *so real* as to commend itself to one's judgement as being the wisest and best thing to do'. He, like the leader writer in the *West Australian*, foresaw that if the Owen Anchorage scheme was adopted, the inner harbour scheme would eventually have to supersede it, but warned that then would arise the question of dealing with vested interests at Owen Anchorage.[5]

At Venn's request, in preparation for the coming debate in parliament, O'Connor prepared a report on the feasibility of siting the harbour works either at the entrance of the Swan River or at Rocky Bay. He demonstrated that such a harbour could be constructed at the entrance 'to meet requirements for some years to come' for £560,000. The cost included excavation of the rock bar, the provision of breakwaters north and south of the entrance, and construction of a wharf 3,850 feet in length. He then estimated the cost of further work for future development, bringing the total cost to £800,000. He explains his reasons for believing that the sand travel arguments had been exaggerated and misunderstood, and concludes his report by referring to the Rocky Bay option. This, he stated, would have some advantages, but these would be counterbalanced by the disadvantages of having to bridge the railway and road twice (or provide road and railway deviations).[6]

All was then set for the debate in parliament in January 1892, when Forrest confidently put his motion:

That this House approves of the scheme of harbour improvement for the port of Fremantle as proposed by the Government, which includes opening a passage through the Success bank into Owen's Anchorage, the construction of a wharf at or near Catherine Point, and a connection by railway from such a wharf to the Customs House and Goods Sheds at Fremantle.[7]

But Forrest misjudged the mood of the House, which was nervous of committing so much loan money to a scheme that had generated a considerable amount of conflicting opinion. The conflict was best summed up by T. F. Quinlan, Member for West Perth:

We want what we have been waiting for for years—for the last 30–40 years and what nature has provided us with in the river. We want this money spent in the proper direction and which will give us a suitable harbour, and encourage vessels to come here, and reduce freights instead of costing as much to get our goods from Owen's Anchorage to Perth as it does to get goods from London to Melbourne.[8]

The Assembly divided and Forrest's motion for Owen Anchorage was defeated. On 19 January 1892, Charles Harper, Member for Beverley, successfully moved the appointment of a Joint Select Committee of both Houses to inquire into 'the Question of Harbour Works…and report what plan would be the best to give secure accommodation to the largest class of ocean-going steamers'.[9]

Thus a momentous decision affecting the future of Western Australia had received a stay of execution, and it was up to O'Connor to convince the committee of the soundness of his recommendations.

The Joint Parliamentary Select Committee met over four days from 22 January, with Venn in the chair. Other members among ten in all were Charles Harper, G. W. Leake, E. T. Hooley and W. E. Marmion. The main witness called was, naturally enough, the Engineer-in-Chief, who was subjected to close questioning on three of those four days. Two weeks before, coincidentally on the day of Forrest's original motion, he had celebrated his 48th birthday. O'Connor was at the height of his powers, confident, thoroughly briefed, and always patient and courteous with his questioners. He gave concise, clear answers that, although in support of his preference for a harbour at the river entrance, never appeared to overstate or embellish the case. It is no exaggeration to say that one of the finest and most successful ports in Australia, which laid the foundation for the colony's subsequent mercantile success, owes its existence to the evidence O'Connor submitted over those three critical days. Some of his answers, quoted here verbatim, show the clarity of his thinking and his grasp of the essential elements of the case.

On the first day of his examination, Monday, 22 January, O'Connor was asked about sand travel, which had been one of Sir John Coode's major objections to a river port. What had made him change his mind?

> Simply from further data I have obtained from people in the locality, and from reports that I had not seen at the time. My apprehension of this sand-travel difficulty is now getting less and less. There seems so little evidence to base it upon. I myself cannot see any tangible evidence of it on the ground, and I cannot find anybody who can show it, or prove it.

Marmion then accused him of pitting local unnamed sources against the evidence of the eminent world authority, Sir John Coode. O'Connor replied:

> I think that is hardly a fair way of putting that question. I did not say that I pitted anyone's evidence against Sir John Coode. I do not think I ought to answer that question at all in that form. I am not aware that Sir John Coode has given any evidence as to the existence of sand-travel; he simply cites certain statements made to him that there was sand-travel. I should not like it to be understood that I am pitting anyone's evidence against Sir John Coode's evidence, nor did I do so.

William Edward Marmion, long-time Member for Fremantle, indefatigable committee man, vocal supporter of Coode's Owen Anchorage plan, and having extensive business interests in Fremantle, would prove a constant irritant to O'Connor and his engineering staff during the building of Fremantle Harbour.

Charles Harper displayed a more measured, independent opinion. Once described as a cold man who 'wants a gallon of blood pumped into him',[10] but nevertheless possessing calm, judicious qualities, Harper asked O'Connor whether he believed that if Sir John Coode were to be made aware that sand travel was infinitesimal, he would then be prepared to recommend a solid structure (a mole) from the ocean. In answer, O'Connor would not be drawn on what Sir John might or might not recommend, but related what Coode recommended at New Plymouth in New Zealand, where the sand travel was much greater than at Fremantle. There, Coode decided that the sand could be eliminated very simply and cheaply by the process of dredging.

To a question about the problem of removing the rock bar at the entrance to the river, O'Connor replied:

> There are three ways of doing it, according to the character of the material to be removed. A great deal of it may be dredged out straight on end by a powerful bucket dredger, especially if we get it sheltered to some extent by the mole. That is one way. Another way would be to use the Lobnitz ram where suitable; and in the next place, where the rock is hard, to blast it, and lift it with Priestman grabs. All rock treated in the last mentioned way would be used directly for the breakwater, and the extra cost met by so utilising it.

At the commencement of the second day of O'Connor's appearance before the committee, Charles Harper subjects O'Connor to an abrasive barrage of questioning, and after raising once again the perceived difference of opinion between O'Connor and Sir John Coode, he asks O'Connor whether he has personally investigated the Owen Anchorage scheme. O'Connor answers:

> Not in detail; but a general view of the chart seems to me to put all these schemes out of court. The five fathom bank outside involves an element of danger for all time, which it would be highly desirable to avoid. If it is possible to get a harbour for a colony which does not involve such a danger it would certainly be preferable.

'Could not that detrimental element be obviated, in fact by dispersing by dynamite?' retorts Harper.

'I do not think so. It is too much exposed', replies O'Connor.

A question put by G. W. Leake—possibly as a friendly opportunity for O'Connor to state his qualifications—asks what experience O'Connor has had in marine engineering 'of this nature'.

O'Connor replies simply by listing his work, first in Ireland, and afterwards his twenty years on marine works, the conservation of rivers, and the building of harbours on the West Coast of New Zealand. He mentions that Sir John Coode approved of his designs for Hokitika Harbour and that O'Connor himself prepared all the data and charts for Coode in designing Greymouth and Westport harbours. He also states that he advised the Government of New Zealand on harbour works at Nelson, New Plymouth, Gisborne and Timaru, where mountain rivers and sand travel were more serious problems.

On his third day of close questioning, O'Connor ranges over his experiences elsewhere, the building of 'island harbours'—that is, harbours connected to the shore by long piers or jetties—and the detailed costs likely to arise from a harbour in the river. In his evidence, he often gives examples of similar harbours elsewhere in the world and shows that he has studied the most telling details carefully, making pertinent comparisons between them and the situation at Fremantle.

Towards the end of what must have been an unpleasant and wearying experience for him, he is asked two decisive questions by his minister, Harry Venn.

'Have you any doubt in your own mind of the ultimate success of this scheme as a completed scheme?'

O'Connor answers simply, 'None whatever'.

Venn, like a lawyer at a criminal trial, expands on his question, emphasizing the gravity of O'Connor's answer and what will rest on it.

'You have no reason to suppose that it will be a failure, and that the colony, after spending a large sum of money on the river scheme, will be driven elsewhere to secure a permanent harbour?'

Thus he shows that the responsibility placed on O'Connor's shoulders is daunting. But O'Connor is unshaken, his answer emphatic.

'I have no reason whatsoever to think so.'[11]

Shortly afterwards, the committee concluded its work. In view of Marmion's antipathy to O'Connor's plan, it is a tribute to Venn's chairmanship, and O'Connor's incontrovertible evidence, that Coode's recommendation was overturned.

This was the first of many parliamentary inquiries and commissions that O'Connor would have to face, and in later years they would become a major distraction from his work, and prove so stressful that they would undoubtedly affect his health and lead to his untimely end. In this instance, however, the inquiry reflected well on him, and increased the respect with which he was regarded generally. As the *West Australian* observed of O'Connor:

> He nearly wrecked all his chances by waiting until the very eve of the final decision before giving expression to his views. But through all this time he was collecting material, examining the various projects, studying the conditions of the ground, bringing his great experience to bear, devising, testing, deciding. Patiently and unhasting he worked out his theories. He was able to put his case so perfectly and unanswerably that conviction followed his evidence.[12]

The Select Committee reported to parliament on 15 February 1892 with a recommendation that O'Connor's plans be accepted. Forrest had no alternative but to drop his Owen Anchorage scheme and speak in favour of the harbour in the river, which he did graciously, paying special tribute to O'Connor's efforts. Charles Harper, who had lately subjected O'Connor to persistent questioning, injected a note of humour into the debate by pointing to the irony in Forrest's change of heart:

> We have been told this evening that the Ministry possesses many virtues, particularly among them the one that when smitten on one cheek, they turn to their opponents the other. For some months past the Government has been in labour, and at last they brought forth a child in the shape of the Owen Anchorage scheme: but after some discussion they repudiated their own child, and took up that of their opponents. If this is not true Christian virtue I do not know what is.[13]

Once parliament had approved of the scheme, and before work could begin in November of 1892, there was much preparation: letting contracts, ordering cranes and trucks and dredgers, and laying the railway from Rocky Bay to Rous Head to convey rocks for the construction of the North Mole. In this enterprise, O'Connor was greatly assisted by his newly appointed chief clerk, Martin Edward Jull, and so began a long and fruitful professional association as well as a close personal friendship.

O'Connor commenced his routine of spending Mondays in Fremantle, working from his home in Quarry Street, or riding his horse, inspecting work in progress and collecting data for his plans and estimates. To men working on the harbour, he became a familiar figure, whether riding around the land-based works, or water-borne in the river on the launch he purchased for his department. 'They came to know him well during those early years as probably no other group of workmen in the colony did.'[14] When Frank Stevens, another recruit from New Zealand, was appointed his private secretary, Stevens would work with O'Connor at 'Park Bungalow' late into the night. The secretary would often sleep in the house ready for an early start the next day. And in those months of preparation, O'Connor could not neglect his responsibilities for the colony's railways. As general manager with wide powers and influence, he introduced the reforms that established his probity in the eyes of most, but sowed the seeds of resentment among privileged contractors that grew and smouldered until the end.

The first conflict erupted in the hot January of 1892, within six months of O'Connor's arrival. After having been awarded the contract for building the

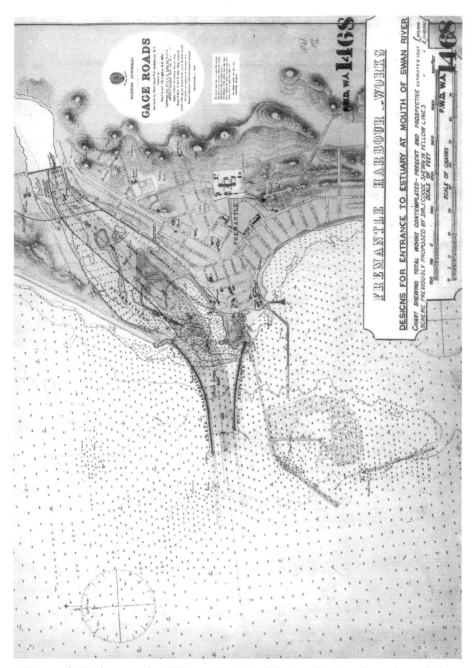

Fremantle Harbour works. O'Connor's original design, showing the moles and the south wharf (Victoria Quay). The rock bar at the entrance is discernible on the chart and so too is the sandy promontory jutting into the centre of what was soon to become the harbour basin. *Original held at SROWA*

Swan River Railway Bridge on the Perth–Bunbury Railway, the contractors, Atkins and Law, contended that longer piles were needed than those shown on the original drawings; that these were never anticipated; and that on enquiry during O'Connor's absence a member of his staff had agreed that the extra expense to Atkins and Law would be covered by the Public Works Department. When O'Connor returned, Venn asked him for his opinion and whether the contract should be amended. But O'Connor was confident that his staff had made no such commitment and that Atkins and Law had deliberately—or unintentionally—misinterpreted what they had been told. He then wrote a masterly juridical analysis on contract law and, at the time, showed remarkable political foresight. In summary, he concluded that the function of the contract was, in fact, to define what had to be done as part and parcel of the contract and within the contract's sum. He saw that the supply of piles of greater length, under certain circumstances, was as much part of the contract and as much within the contract's sum as any other requirement.[15]

The dispute continued, and in another memorandum, dated 4 March, O'Connor reminded the minister that the department was just beginning a large scheme of public works and 'as we begin so probably shall we have to go on'. He diplomatically avoided accusing local contractors and merely instanced that

> in some colonies the influence they bring to bear nearly always succeeds in relieving them of the penalties for delay or higher costs either by petition to the House or other means.[16]

He then shows compassion for the poorer contractors who had no influence among the legislators, and states that engineers had hesitated in the past to recommend penalties, knowing that other contractors, who had friends in power, were regularly compensated. He ends by arguing that unless there were very strong reasons for the amendment of contracts, penalties should be rigidly enforced and that such a policy would be 'to the advantage of all contractors who will make provision accordingly in their contracts and the work will go along promptly and satisfactorily'. In reply, Venn stated that he was 'entirely in accord'.[17]

Western Australia at that time was governed, and its business administered, by a relatively small number of powerful families—not so much an old boys' network as an old settlers' network—and there was suspicion of outsiders who brought in new ideas and confident professionalism. Fortunately for O'Connor, his worth was recognized and valued by the Premier and his immediate circle, and perhaps grudgingly by other members of the oligarchy. But those who were jealous of his success and judged his undoubted power as autocratic would let no opportunity pass for censuring him. As already mentioned, one of the most vocal was the Premier's brother, Alexander Forrest.

That O'Connor's methods and self-confidence, to some extent, inflamed later criticism has to be admitted. Like most educated professionals, imbued with Victorian self-confidence, he saw clearly how his work should be done properly. He was evidently impatient with the colonial public service mentality. He cut corners, and was never one to suffer interfering amateurs, of which there were many in parliament. And yet there is much evidence that his professional colleagues admired him and were devoted to him. His secretary, Frank Stevens, thought him 'a genius, possessing extraordinary foresight'.[18] His first biographer, who had the advantage of talking with men who had worked with him, has written that he was

> neither dogmatic nor arrogant [and] possessed an extraordinary capacity to win the interest and cooperation of men whose experience enabled them to throw light upon his plans. Never did he fail to acknowledge information offered.[19]

His daughter Kathleen wrote: 'I have known him thank someone for a most simple suggestion, and, childlike, wondered at the time [why] he had not thought of it himself'.[20]

O'Connor's character was complex. He was reserved, much dedicated to his profession, and a very private man, ill at ease in the corridors of political power. There is some reason to believe that, in spite of outward appearances to the contrary, he suffered occasional bouts of depression, and this tendency is explored in later chapters.

Just as public opinion had pressed for the construction of a port at Fremantle, equally vocal and insistent were the calls on the government to extend the Eastern line beyond York to Southern Cross, to service the new goldfields of the Yilgarn. Gold seekers were flooding into the colony, arriving at both Albany and Fremantle, their final destination 300 miles inland at Southern Cross. As early as 1890, John Forrest, who as Commissioner of Crown Lands had visited the area, advised building the railway. Now, as Premier, he was saddled with fulfilling his own recommendations.

Naturally, Forrest turned to his Engineer-in-Chief and General Manager of Railways to deal with the practicalities, and make it all possible.

16

The Murchison goldfields in central Western Australia were named after that unremarkable but capricious river that is a dry creek bed for most of the year and occasionally a raging flood during the short wet season.

No such river, wet or dry, inspires the Yilgarn in the central south of the State; it takes its name from an Aboriginal word for the white quartz found in abundance around those parts. Rectangular in shape, on the fringe of desert country, the Yilgarn stretches north and south for about 200 miles, having the town of Southern Cross—or simply 'the Cross'—at its centre. Gold was discovered there in 1887 and a workable field was proclaimed a year later. In 1891, O'Connor's first year in the colony, the Yilgarn fields produced 12,833 ounces of alluvial gold, then worth close to £49,000. This figure would nearly double in the following year, and would yet rise to 75,000 ounces (worth £287,829) in 1893. Such riches attracted a motley stream of prospectors from eastern Australia and from as far away as New Zealand and Europe. Those travelling by way of Fremantle in the west, and Albany in the south, came by train as far as York or Beverley, but neither of these towns was more than a quarter of the total distance. From there, the only way eastwards was on foot or by wagon, 160 miles over a stony track. The land was mostly featureless, a dusty plain broken by low hills covered with monotonous salmon gums and mulga. Those who could afford to travelled by Cobb and Co. coach; others with less money paid 1 shilling per pound for their swag and provisions to be carried by horse or camel, while the owner, unencumbered, followed behind on foot. Most had to walk, pushing their own barrow-load of pots and pans, water bags and assorted equipment, or humping their swag on their backs. There was little groundwater, mostly limited to a few Aboriginal wells. A tank of 1 million gallons' capacity had been dug south of the Cross, but the scant rainfall had

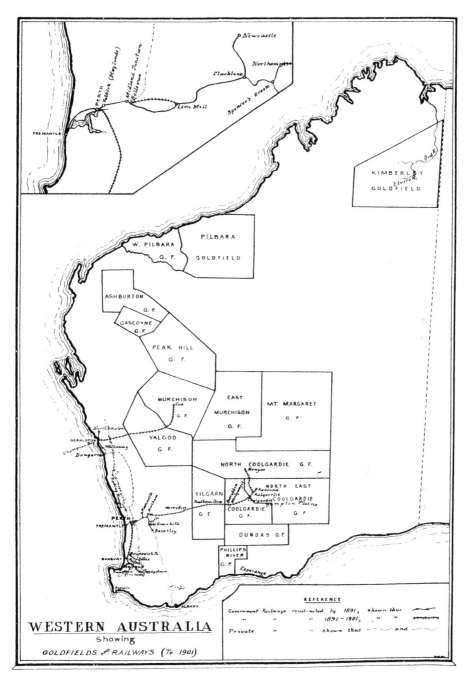

Goldfields and railways in Western Australia, 1901.

Map by Nancy McMillan (1978)

been unable to fill it. Notwithstanding these obstacles, the lure of gold and the prospect of instant riches were generally enough to inspire optimism in the hearts of the trekkers, although many of those who set out on the journey in high spirits did not complete the distance, as Albert Gaston recounts in one of the few published personal records of those times:

> After a bit to eat and a drink of tea, we unrolled our swags and slept beneath the stars till morning. While I was boiling the billy for breakfast my mate investigated a blistered heel. He seemed inclined to abandon the trip. However I persuaded him to keep going and assured him that by cutting a hole in his boot he would be all right. We set out on our second day's tramp, but now the going was very slow.
>
> We were unable to reach water that night and had to do with little drops in our bags. My mate said he would go no further. He could not remain there without water so I carried his swag on top of my own. A sorry pair we must have looked.
>
> At length we arrived at a well and camped for the day, or what was left of it. The next day Jim was no better, so I was forced to go on alone. I afterwards learned he had returned to York and condemned all swags and the fools who carried them.[1]

Gaston travelled with a gallon of water in his bag, and well before he arrived at the fields he was using his last pint for a billy of tea and complaining of feeling lonely, weary and hungry.

Three years before the legendary gold discoveries of Bayley and Ford at Coolgardie in September 1892, it had been widely recognized that a railway and telegraph line, and a reliable supply of water, were urgently needed if the gold-fields were to develop and disease and death to be prevented.[2] It was on this account, and probably at the prompting of his minister, that O'Connor decided to see the country for himself and assess conditions for an extension of the Eastern line as far as Southern Cross. 'The chief object of his visit', stated the *West Australian* of 26 August 1892, 'is the inspection of the railway route with a view to facilitate the calling of tenders. Mr O'Connor expects to be absent twenty days.'

He set out on 31 August, coincidentally two weeks before Arthur Bayley's dramatic arrival back in town from Coolgardie with more than 500 ounces of gold in his saddle bag. Bayley's find would lead to the greatest rush of the century and the emergence of Coolgardie as the goldfields capital of the West. O'Connor's surviving notebook of the Yilgarn trip is frustratingly terse, limited to dates of arriving at towns along the route, a few jottings, and occasionally a reminder to himself of work to be completed. Before setting out, he lists his considerable personal belongings not to be forgotten: his coat, valise, umbrella (did he expect rain?), tin box, dressing bag, dispatch case and office bag. He arrived in Toodyay (then called Newcastle) on 1 September, and Northam the next day. Although he does not say, we can assume he travelled first by train and then by hired trap from York or Northam. He stopped overnight at settlements on route—Tammin on the 7th, Merredin on the 11th and Burracoppin on the 12th—arriving at Southern Cross on 14 September. He was three days at the Cross, a thriving town at that time, with shops, hotels, banks and office. While there, he was entertained by the town council and ratepayers. In his speech, O'Connor confirmed that the railway survey had been economically and expeditiously completed. He apologized for the delay and suggested that it might take up to two years for completion, but he would recommend that the railway be opened as it proceeded, thus shortening the distance to be travelled by road. The newspaper report of 26 September stated that Mr O'Connor 'left a very pleasing impression upon those who had the privilege of meeting him'.[3]

The Engineer-in-Chief may have learned of Arthur Bayley's gold find while he was in Southern Cross. If his diary is correct, he was still in the township on that eventful day, 16 September, when Bayley 'sauntered into the bank in an unconcerned manner' and handed over 554 ounces of gold then worth around £2,000.[4] He had knocked it out of a reef with a hatchet, and although there was evidence of much more where that had come from he had hastened into town to register his claim. With O'Connor's experience of similar finds on the New Zealand West Coast, the significance of this latest find, and the changes it would bring to the colony, could not have been lost on him. Even within hours of Bayley's registering his Reward Claim, 'the excitement rose to fever pitch' and prospectors, on receiving the news, started heading for the new fields.[5]

With this event, the new railway that O'Connor was then planning as far as Southern Cross would, even before the track was laid, bring demands for an extension 120 miles eastwards to Coolgardie. The Engineer-in-Chief, or 'The Chief' as he was more popularly known, was back in York on 21 September and in Fremantle on the 22nd. The last two jottings in his notebook refer to the cost of a cab at Fremantle (3 shillings) and a reminder about sending railway plans to Southern Cross.[6]

On returning to his office, he immersed himself once again in preparations for the work on Fremantle Harbour and found in his absence that there had occurred the first of many subsequent protest meetings in the town, deploring his recommendation for moving the railway workshops from the river foreshore to a site in Midland. Local interests feared that unemployment and reduced trade would result. But it was clear to O'Connor not only that the existing workshops were inadequate and lacked facilities and experienced staff but that their present position would obstruct his plans for the new harbour wharves. On O'Connor's recommendation, the locomotive superintendent of the Victorian Railways, A. D. Smith, a railway engineer of long experience, had been engaged to advise the government on the future of the workshops and provide estimates of the cost of relocating them. Although his report confirmed O'Connor's ideas, in a debate in parliament in December 1892 a motion to act on the report was defeated, the Premier himself voting against any further action.[7] This, however, was only the first skirmish in a long battle. The conflict over the workshops would erupt again in the ensuing years as the harbour work progressed and relocation became more urgent. Local traders and businessmen, unable to foresee the compensating benefits that would flow from the construction of a new port, with its increased shipping, subjected O'Connor to some of the bitterest attacks of his career in Western Australia.

Perhaps the conflict was forgotten for a brief time in the excitement of the official inauguration of the new harbour works on 16 November 1892. Lady Robinson, wife of the Governor, performed the opening ceremony by tipping the first truck of stone brought from Rocky Bay to begin forming the North Mole. Before all the dignitaries present and a huge crowd, Sir John Forrest paid generous tribute to C. Y. O'Connor, 'an able, brave, and self-reliant man'. He admitted that at first he had not supported O'Connor's plan for the harbour, believing it too costly and too risky, but that he had been won over. Now he looked forward to the time when Fremantle would 'become the Brindisi of Australia…I give all credit to the Engineer-in-Chief'.[8] Forrest remained a firm admirer of O'Connor and always spoke well of him in public, although he was often slow to support O'Connor's plans in parliament—the river harbour and the railway workshops being just two examples.

In the heat of the New Year 1893, O'Connor celebrated his 50th birthday, and at the same time the family faced a disruption to their domestic routine. On 11 January, the *Inquirer* announced the return from overseas of Colonial Surgeon Dr Calvert Barnett and his intention of reoccupying 'Park Bungalow'. The O'Connors rented 'Yeldam House' for a few months and then took up residence in 'Plympton House', owned by William Pearse. Of the same style and proportions as 'Park Bungalow' and almost in sight of it, 'Plympton House' overlooked Beach Street and the river from an elevated position. It had the advantage of adequate stabling and outhouses on three lots: Lots 261 and 262 fronting Beach Street and Lot 307 on Cantonment Street. Those searching for the house today will not find it, but a clue to its position is the old iceworks, which was later built on the site, next to the large furniture storage building known as Fort Knox. More demolition and rebuilding planned in this area will hide the exact location forever. But looking at Beach Street today, behind the Flying Angel sailors club, it is not hard to imagine how O'Connor must have rode on his horse down the slight limestone incline from the stables, across Beach Street, over the profusion of goods railway lines (now simplified and raised on an embankment), to the river bank with its reeds and shallows, which would shortly be reclaimed and become part of the wharf system. Years later, his daughter Kathleen remembered how the Irish stable boys would conduct the girls down to the river for a swim or a sail.[9]

In those early months of construction of the harbour, O'Connor set up an office in an old railway shed. As the number of staff working on the harbour increased, so too did allegations that the Engineer-in-Chief was appointing his old New Zealand cronies in preference to West Australians. Venn came to his rescue in parliament and stated that, unknown to O'Connor, he himself had made investigations into the Fremantle works and found that New Zealanders, who were said to be so numerous because of the preferences shown to them, were really in a minority to 'an enormous extent'. The return showed that the clamour that had been raised was altogether unfounded.[10]

Throughout 1893, the appearance of Fremantle—at least around the river entrance—began, imperceptibly at first, to change. The North Mole, so familiar today to countless migrants, seamen of all countries, and round-the-world yachtsmen and women, was gradually taking shape. All day long (and later, during the night when shift work was introduced), steam engines shunted wagons loaded with blocks of stone weighing from 1 to 30 tons. These were emptied into the water, increasing the length of the pier. Like the movement of the hands of a clock, the work proceeded so slowly that only the result, and not the movement, could be perceived. At first, the rock came from Rocky Bay, a short distance up river, but this was found to be inferior stone and new supplies had to be brought all the way down on the Eastern line from a quarry in the hills. This mole structure, known as Pierres Perdues (literally, waste stones), was in contrast to the more formal concrete groin structure common in Europe. The width of the flat top of the new mole, where the railway rested, was 60 feet, and the structure was 17¾ feet above low water. A rough, rubble parapet wall on the seaward side provided added protection from the weather.

Once the mole had advanced far enough to afford some protection from the send of the sea, dredging and rock blasting could begin. O'Connor described this as 'really the most important work of the whole scheme, as, on its being satisfactorily done, depended the success of the undertaking'.[11] Fremantle residents must have viewed with curiosity the appearance of strange trestle stages in the middle of the shallow river entrance. From those stages, holes 3 inches in diameter and spaced 10 feet apart were drilled into the rock below to a depth of 33 feet. As each hole was made, it was charged with a watertight tin tube reaching from the bottom of the hole to above the water level, and these in turn were plugged with gelignite and dynamite. The trestles were removed and the charges exploded by means of time fuses. The trestles were then set up a distance away and the process was repeated again and again. Because of complaints from the townspeople about vibration from the explosions, the amount of dynamite fired at any one time was limited to 50 pounds, 'which considerably impeded the progress of the work'.[12] Bucket dredgers would follow the blasting and these, according to O'Connor, lifted over a million cubic yards of rock, much of it used for the North and South moles. During all this work, O'Connor noted detailed costs—for example, the value of the rock per ton dredged from the river and put to good use, and the savings thus made. There were bound to be unforeseen expenses, one of them the capsizing of one of the Priestman bucket dredgers in May and the delay and expense of bringing it to the surface.

Whether O'Connor enjoyed official engagements is debatable, but his inherent courtesy and Victorian sense of social propriety ensured that he was always present at vice-regal and other official functions. For example, his work was left behind in Fremantle when he attended, with his wife and daughter Aileen, yet another reception at Government House on 16 May 1893. But he was absent from the official opening of the Perth–Bunbury Railway on the 24th. This may be explained perhaps by a later report that he had been ill and confined to his house for several days.[13] His illness was probably influenza brought on by the inevitable change in the weather at that time of the year exacerbated by his continual work inspections and disregard of the conditions on the water at Fremantle.

Influenza it may have been, but the symptoms surely would have caused his family momentary concern, coming as they did on top of an outbreak of typhoid in the colony. There were two reported cases in Fremantle. The papers described the few cases prematurely as 'an epidemic' and demanded action from the health authorities. Alarmist as the newspaper reports and editorials were at that time, the appearance of the disease was the beginning of what became a notorious typhoid plague bringing many deaths and much illness mainly—though not exclusively—in the goldfields. The shortage of water there, and the consequent primitive living conditions, intensified calls for a plentiful, clean water supply.

Fortunately, O'Connor and his family were not among the victims. 'The Chief' was back at work on the last day of the month.

17

WITH THE ARRIVAL of the year 1894—O'Connor's third year as Engineer-in-Chief—Western Australia was entering upon a remarkable period of economic growth, of optimism and of prosperity. That this was due principally to the dramatic gold discoveries in the Yilgarn and Coolgardie fields is beyond question; the seemingly ever-increasing wealth on those goldfields, together with the establishment of new companies, increased overseas investment and confidence, and the influx of migrants, would carry the colony forward at a giddy pace throughout the remainder of the decade.

And yet gold discoveries, while undoubtedly the key to the colony's transformation from Cinderella to the desirable bride envisaged in John Boyle O'Reilly's poem, would not alone have been sufficient. An essential ingredient was the union of C. Y. O'Connor's talents and the statesmanship of John Forrest, supported by similarly idealistic members of his Cabinet. Great turning points in history need leaders of genius to shape them towards the general good (or sometimes, alas, to the opposite), and always it has been a moot point whether great leaders predetermine events or whether events occur at random and leaders arise and grow in stature in their wake. Certainly, O'Connor played a pivotal role, hard to exaggerate, in the development of Western Australia at that time. He 'changed the order of things' to the extent that his planning and his foresight would leave an indelible mark on the map of the State for the next hundred years.

Late in January 1894, we find him travelling once again to Southern Cross in company with his minister, Harry Venn. In the scorching heat of that month, it must have been an arduous duty. The railway line had made progress but had reached only as far as Burracoppin. From there, Cobb and Co. coaches ran twice a week the remaining 52 miles to Southern Cross, where the lack of water was so grave that the frequency of horse-drawn traffic had been restricted in order to give the limited number of soaks a chance to refill. There would have been no air-conditioning and no cool showers at Booth's Hotel, where the ministerial party was entertained at a banquet, 'the largest of its kind ever held at Southern Cross'.[1] Toasts were drunk to Venn and to O'Connor, who were seated beside the mayor, Captain Oats, and the much respected warden, John Michael Finnerty. Uppermost in the speech of Captain Oats was the plea to the government for an improved and adequate water supply. 'He concluded by proposing the health of Mr Venn which was drunk in bumpers of champagne with much enthusiasm, and musical honours.' In reply, Venn said that he was much gratified at the kind and flattering reception that was accorded him. He spoke of the government's commitment to the goldfields and defended the policy of borrowing money for public works, a policy that had been criticized by candidates in the Yilgarn electorate. With regard to water supply, he said:

> the Government had secured men of the highest attainments in connection with water conservation…[they] would leave no stone unturned to solve the difficulty, and secure to the fields what was most wanted at present—water.

Venn himself was a gentleman farmer from the well-watered pastures of the Bunbury hinterland, noted for his rotundity, his ruddy complexion and his panama hats. His speech provoked loud applause and afterwards glasses were raised for a toast to the Engineer-in-Chief. O'Connor responded by thanking them and saying that it was the first occasion his health had been drunk at Southern Cross. He then reviewed the progress of public works undertaken so far—the extension of government railways, and work in connection with Fremantle Harbour. He alluded to the criticism that too many New Zealanders had been employed in these works but said that 'the facts showed that out of 173 men employed, only 11 had come from New Zealand'.

A considerable number of toasts seemed to have been consumed, as the *Inquirer* described, 'with great enthusiasm', and the evening ended with songs and recitations that 'brought to a close a most enjoyable function'.

Back in Perth, O'Connor was embroiled in a public controversy arising from accusations of improper conduct alleged in the Auditor-General's report to parliament of 1892. He had been accused in the report, by implication, of having lent ironwork and other goods illegally and without authority to one of his contractors engaged on building the Jarrahdale Railway. The Parliamentary Committee investigating the matter had found that the allegations were entirely false and that O'Connor had acted properly and legally. The government accepted the findings of the committee and ignored the accusations. There the matter might have rested. But this was insufficient for O'Connor; he wanted a public retraction:

> I feel constrained as a duty to myself, as well as to the profession to which I have the honour to belong, to ask the Government to make some distinct and public announcement of the fact that this charge is entirely unwarranted by the facts of the case.

The correspondence, including an angry letter from O'Connor giving full details of the case, was printed in the *Inquirer* of 16 February under the provocative heading 'The Auditor-General and the Engineer-in-Chief'. Printed below it was a response from the Premier that must have mollified O'Connor: 'I have very great pleasure in recording', Forrest wrote

> that there were no grounds whatever in connection with the issue of the iron work, wheels and axles to Mr Neil McNeil or for any implication of improper conduct on your part and that you had full authority to issue them.[2]

The incident, which may seem trivial in retrospect, is nevertheless a useful measure of O'Connor's proud character. The committee had cleared him of improper conduct but his honour had been questioned in an official document. Not for the last time in Western Australia, he acted like the duellists of the previous century. Having received what he saw as an insult, he demanded not just an apology but public satisfaction. Seven years later, it would be this same pride and inflexible sense of honour, born of his evangelical upbringing and his lifelong dedication to his profession, that would present him with his ultimate nemesis.

Although the shortage of water in the goldfields had been well recognized prior to the Venn and O'Connor visit to Southern Cross in January–February 1894, from that date forward the situation became a primary concern for the Engineer-in-Chief, as much his responsibility as the harbour works and managing the railways. Scarcely a day passed without the water shortage being alluded to in the daily papers, either in the news columns or the editorials. 'The year 1894 will be long remembered as one of exceptional dryness, and at the time was characterised as one of the worst in the annals of the colony.'[3] In May, it was announced by Warden Finnerty in Southern Cross that only four teams every twenty-four hours are to get permits to travel to Coolgardie, and no permits will be given to teams loaded with men's swags, provisions only being allowed to be taken up.[4]

The message for the government was clear: solving the water problem in the goldfields must be a priority, or the new-found wealth of the colony could be jeopardized. O'Connor set up a separate sub-department, the Goldfields Water Supply Department, within Public Works, with a resident engineer, Charles Jobson, on the spot reporting to the Chief. Jobson's reports forwarded daily were invariably printed in the newspapers, eagerly looked for by the mining community and investors, in much the same way as cyclone or storm warnings are today. They commenced on 27 April under the heading 'The Water Difficulty (By Telegraph)'. On 4 May, the *Inquirer* stated that Southern Cross was in a very bad way for water:

> It is now absolutely necessary that a water train be sent daily. The Government condenser is quite inadequate to meet the demands of the population here. Horses are dying, and the distress is beginning to be felt. Four permits are issued every day to teams going to Coolgardie as the temporary stoppage of teams has allowed the soaks to make a little more water but still there is a 48 miles stage waterless.

O'Connor set out for Southern Cross on 3 May, the second visit within three months. His purpose, the *Inquirer* reported, was 'to ascertain what are the best steps to be taken to secure a water supply to the inhabitants'.[5] Various impractical schemes for solving the problem were being put forward and debated in the newspapers—anything from digging canals connecting with salt lakes (ignoring the fact that the lakes were invariably dry), or building a special horse-drawn tramway to bring water trucks from Northam, or, closer to reality, constructing a pipeline from the Esperance coast. Although a water pipeline from the west coast had not, at that time, been seriously considered by the government, there had been two suggestions put forward for pumping water from the Avon Valley: the first published in the *West Australian* on 8 March 1894; and the second, an application from a private investor, in August of the same year. Both sought the right to construct a pipeline along the railway line from Northam to Coolgardie. Although neither of these proposals, which are discussed in more detail in chapter 18, materialized, they bore certain fruit by raising in the public mind the possibility of solving the water problem by the method that was eventually adopted.

But perhaps the honour of being the first to propose the idea of a pipeline belongs to a professional engineer from Victoria, E. S. Heath, who wrote to the Premier as early as November 1893. He claimed that 100 miles of 3–7 inch pipes could be laid from a water source within sixty days, and he cited as examples the pumping of oil in Pennsylvania and his own work in supplying Kitchener's forces with water in the Sudan. He made no mention of using storage dams in the hills east of Perth.[6] The Premier directed the letter to O'Connor's office and so E. S. Heath may have been the one to sow the seed from which the Coolgardie Goldfields Water Supply Scheme, as it was originally called, eventually took root. If so, that seed lay dormant in the department files for a lengthy period. The only water then obtainable in the goldfields area, apart from supplies brought laboriously and expensively from the west, was derived from tanks very occasionally filled by rain; from rock pools; and from condensers that distilled the extremely salty fluid obtainable in wells and mine shafts. The average rainfall amounted to as little as 3½ inches per year, and the surface soil was so porous, flat and saline that natural freshwater sources were practically unknown. It had been argued that water lay under the ground at considerable depth, but wherever expensive bores had been sunk—one to a depth of 3,000 feet through granite rock—results had been disappointing.

The *Inquirer* of 11 May described the situation at Southern Cross as a 'water famine' and commended O'Connor for journeying there again 'to see how the working of the tanks and dams and improving the soaks is proceeding'. And a week later, the same paper printed a long report from O'Connor about conditions on the fields and recommending strict policing of the permit system. He referred to the conflicting newspaper reports about water availability along the route. He cited one traveller, a Mr Cohen, who reported that he could get no water from the wells, but this was denied by a Mr de Baun in the same paper, who found water at several places. It is possible, O'Connor wrote, that both men were right. A soak that was capable of supplying a drink to each of forty horses, he explained, would appear dry to the next team that arrived there if the time elapsed between the two teams arriving was insufficient to allow the soak to refill. The solution, he concluded, was one of strictly adhering to the permit system. He advised that if men without permits to travel took water, they should be prosecuted.

As important as finding water for domestic and mining requirements was, water for the steam engines on the railway to Southern Cross and beyond was of more pressing concern to O'Connor. One of his trains could haul thirty wagons, each carrying a 1,200 gallon tank, a total of 36,000 gallons, but on the return journey the engine itself would need to consume slightly more than half of this amount of water.[7] If water could be stored, or a sufficient supply found beside the rail tracks, wagons would thereby be freed for increased freight, and operations would prove more profitable. In September 1893, O'Connor had sent one of his assistant engineers, William Shields, to assess the country for possible water either side of the railway, but Shields returned with confirmation that no such supplies could be found. He recommended four large tanks be constructed at Cunderdin, Kellerberrin, Merredin and Parkers Road, and by 30 June 1895 the tanks contained a useable supply of water.[8]

As the winter of 1894 merged into spring, the daily reports from Jobson and others on the conditions in the goldfields worsened. Little winter rain had fallen, and by the end of September the weather was hot again, with dust storms adding to the discomfort. A dispatch described how pitiable was the sight of horses in the centre of Coolgardie, tottering with weakness for want of water, and reported that the publicans of the town, unable to supply travellers with water, were closing the boarding and lodging parts of their premises.[9]

In November 1894, O'Connor had other matters to distract him. Opposition to his proposals to move the railway workshops from beside the river at Fremantle, and rebuild them—possibly at Midland—had never entirely subsided. Too much was at stake for local businesses in Fremantle. When the matter was debated in the Assembly, Alexander Forrest, a vocal opponent of the move, used the occasion to cast another slur on the Engineer-in-Chief: 'An able officer', he said, 'but an expensive official, and if we went on as we are going, in ten years the colony could then be handed over to that gentleman'.[10] On an amendment deleting reference to the site at Midland, the motion 'that the railway workshops should be removed to a more advantageous site' was passed by eighteen votes to ten. But Alexander Forrest, then mayor of Perth, had not finished with O'Connor. At a protest meeting in the Fremantle Town Hall the following March, he was one of the main speakers. He said that he would rather take the opinion of men of commonsense than get engineering advice. He admitted that there were large vested interests in keeping the workshops in Fremantle and therefore the government should respect those interests. 'It would be far better for the Government to get rid of the Engineer-in-Chief, rather than ruin half of Fremantle.'[11] A voice in the crowd shouted, 'And the whole colony!'—but Forrest replied, to laughter, that he did not think it could ruin the whole colony.

The antagonism between the Premier's brother and O'Connor may have been mutual. In a debate in the Assembly on whether a Commission of Inquiry into the Civil Service should be discontinued—it being argued that it was a waste of the various departments' time and getting nowhere—Alexander Forrest argued strongly in favour of continuing the inquiry, having his sights set firmly on O'Connor's department. 'We know that the Engineer-in-Chief represents and rules not only the Works and Railways but the whole Colony', he said.[12] He then went on to accuse O'Connor of being unapproachable, of being above the minister. 'Why should the whole Colony be ruled by one man, however clever he may be? How was it that New Zealand had let this man go?' he asked. He then suggested, as mentioned in chapter 14, that inquiries in New Zealand had led him to believe that O'Connor had ruled his ministers in that country, too. Neither was Alexander Forrest, MLA, famous explorer, businessman and mayor of Perth, content to leave the matter there; he was clearly representing the views of the aggrieved contractors who bore a grudge against O'Connor and his methods. He cited a case when he had questioned one of O'Connor's contracts in the House, arguing that O'Connor had not chosen the lowest estimate for supplying railways sleepers.

> Some four or five months afterwards I happened to meet this man who rules this country. I went up to speak to him, like one citizen might to another. And he told me, 'Unless you apologise for your conduct in the House (or words to that effect) I do not wish to know you.' I at once said that I thought his reply impertinent.[13]

Again we observe an instance of O'Connor's honour being impugned, and his acute sense of injury and his demand for retraction. That Alexander Forrest's attacks were maliciously wounding to O'Connor there can be no doubt, but a less intransigent reaction by O'Connor would have gone some way to mend broken fences. Even R. F. Sholl, who sprang to the defence of O'Connor in the House—'Everyone acknowledges that he does not spare himself in any way, and that he is a very hard-working official and that he does his work well'—ended by adding:

> It would, perhaps, be much more pleasant for himself if he tried to be a little more conciliatory with contractors and others, but I do not think that the country would benefit by it.[14]

It is clear that O'Connor, by that time, had a daunting amount of work on his hands and faced pressing demands from several directions at any one time. Today we would say that he was 'working under intense pressure' and expect him to be on tranquillizers. But O'Connor seemed to thrive on hard work, his armour being his single-minded dedication and professional self-confidence, which often resulted in a perceived dictatorial manner. Like many professional men of his time who were constantly under pressure, he drank spirits liberally, perhaps more as a corroborant than a social necessity. Although this habit did not affect his powers, his health was to bear some long-term scars—as is discussed in the Epilogue.

It seems that most reasonable people agreed with Sholl's and the Premier's defence of O'Connor in parliament and were content to judge him on the results of his work—and on this basis there was much congratulation and general satisfaction. Leading with congratulations was the *West Australian*, which announced on 7 February 1895 that, almost unnoticed by the general public, the North Mole of the new Fremantle Harbour works had been completed to the length originally planned by the Engineer-in-Chief. The structure had been carried out to sea to a distance of 2,934 feet at a cost of £59,762, which was £20,687 less than O'Connor's original estimate. The report added that the Engineer-in-Chief now proposed to lengthen the mole still further until the sum allotted for the work had been spent. Work on the South Mole was also well in progress, the stone being used coming mainly from the levelling of Arthur Head.

Any pleasure or celebration at the passing of this milestone was dampened by a sudden and tragic accident at the quarry at Rocky Bay, which claimed the life of John Irvine, then officer in charge of O'Connor's harbour works. Also critically injured was the ganger, Timothy Yorke. On Thursday, 7 February (the day of the report in the *West Australian*), Irvine had been testing a new kind of explosive which, it was claimed, was superior to dynamite and cheaper (ironically, made in and imported from New Zealand). A premature explosion injured Irvine's spine and face, and although he survived for a few hours, he died in great pain the following morning.

It was said to be a gloomy day in Fremantle the day following the accident.[15] Work was entirely suspended at the quarries, and flags were flown at half-mast in the town and on the ships in the vicinity. For O'Connor, this must have been a personal as well as professional loss. John Irvine, aged 51, had been brought from New Zealand, previously having worked with O'Connor on the harbours at Hokitika and Westport. The funeral in Fremantle on the Saturday was attended by around 1,500 people, 500 of them employees on the harbour works. Venn, the minister, was present; O'Connor himself was one of the pallbearers, along with his second-in-command, Arthur Dillon Bell. Did O'Connor remember, we wonder, the other tragic accident eleven years previously in New Zealand when his friend and colleague Emanuel Rawlings died from work injuries? In both cases, in those days, there was no provision for workers' compensation or insurance. In Irvine's case, a public collection was taken up to present to the grieving widow and her family. Although the subsequent inquest found that no blame should be attached to any individual, that it was 'accidental death', there remains at this distance in time a suspicion that the pressure to cut costs and the absence of adequate safety procedures were more than likely the cause.

In Sholl's defence of O'Connor in parliament, he had alluded to the difficulty people found in catching up with the Engineer-in-Chief on account of the demands of his large department and the necessity for him to travel constantly all over the country. This was no polite excuse on Sholl's part but based on solid evidence. In early March 1895, O'Connor was off again on his travels, paying his fourth visit to the Yilgarn. His main business was inspecting tanks and bores along the railway route, which now reached Southern Cross. On his way back, he broke his journey, diverting south to Albany, where he remained for ten days inspecting the harbour and advising on design and dredging works. He was back in Perth again on the 27th of the month.

While O'Connor had been away in Albany, yet another meeting had taken place in Fremantle Town Hall, protesting against his proposal to relocate the railway workshops in Fremantle to a new site in Midland Junction. His name was much in the news. Those who opposed the relocation saw O'Connor as the villain, and no one was more eager to do this and speak on the subject than prominent citizen and member of parliament Alexander Forrest. By this time, O'Connor must have been unaffected though wearied by the ill-informed opposition that had been raging for well over three years— opposition to what he saw as a logical development, essential if the new harbour were to be a success. The *Inquirer* (one of whose leading shareholders was Alexander Forrest) had consistently led the attack, arguing that it was O'Connor's duty to find a suitable site close to Fremantle so that local business interests would be protected. If the workshops were to be removed any distance, it opined

> Such a wrench in the business interests in the town and also in individual interests…would not be compensated for by any of the alluring prospects held out to the people of Fremantle by the suspicious sophistry of a section of the Press.[16]

This was a dart aimed at its rival, the *West Australian*, which broadly supported O'Connor and the removal of the workshops. And in a more pointed reference to O'Connor, the *Inquirer* had claimed:

It is plain to every impartial observer, that the railway authorities are more concerned to achieve a reputation as railway financiers than as railway managers and so long as they can show on paper improved financial returns it matters little to them at what cost and annoyance to the public.[17]

The noisy opposition in Fremantle to the removal of the workshops culminated in an unprecedented demonstration and deputation to the Premier on Wednesday, 20 March 1895. Four hundred citizens led by the mayor of Fremantle, prominent businessman W. F. Samson, and including members of parliament W. S. Pearse, Elias Solomon and, not surprisingly because of his business interests in Fremantle, the mayor of Perth, Alexander Forrest, marched through the streets in procession to the railway station. They all boarded a special train for Perth, and resumed their march up William Street and St Georges Terrace to the Premier's office. W. E. Marmion, then Minister for Mines and Member for Fremantle, introduced the deputation to the Premier with regret, because 'the reason for the deputation was one that affected the interests, to a very great extent, of the whole of the people of the colony'. The arguments were put forcibly, not that the workshops should not be removed from their present site but that they should remain in or within easy reach of Fremantle. The Premier's brother and others favoured a site at Rocky Bay now that it had been cleared by quarry work, but the Premier reminded them that the Engineer-in-Chief had stated that Rocky Bay would not be suitable. The Premier tried his best to be conciliatory but he was firm and repeated more than once that he 'took his stand on O'Connor's opinion'. The deputation left without receiving any apparent satisfaction.[18]

O'Connor himself was on record as stating that although he wanted the workshops removed from their present site, he did not really care, from an engineering point of view, where the new workshops were located, so long as there was sufficient space to work in. It was up to the government, he said, to choose which site to use, based on the options available. He then defended the Midland Junction site (already purchased by the government) owing to its level character and the considerably lower costs involved there, in comparison to other sites that had been suggested.

The concerned citizens of Fremantle made yet another valiant attempt to argue that the workshops should remain in the port vicinity by inviting a government party, including the Premier and the Engineer-in-Chief, to inspect two alternative sites on Monday, 19 August. They hoped that by confronting O'Connor in company with John Forrest, they could induce him at least to endorse one of the alternatives put forward. Unfortunately, the day proved one of the wettest of the winter, and the blinding rain and dismal aspect restricted the tour, adding to the futility of their efforts and the general discomfort of everyone concerned. After sheltering, huddled in a ganger's shed, until the rain eased somewhat, the party first inspected Rocky Bay. There, the *West Australian* reported, 'the Engineer-in-Chief had to bear the whole brunt of the attack'.[19] He was asked by the Fremantle-ites whether the ground (cleared by quarrying stone for the moles) was sufficient for the workshops. O'Connor replied, 'No. We should want a hundred acres'. He was then accused of changing his mind, having stated earlier that he required only 20 acres. But O'Connor replied that if he had mentioned 20 acres—and he doubted this—it was because he did not then have in his possession all the relevant information. The arguments went back and forth until the Fremantle contingent grumbled that 'There was no arguing with Mr O'Connor. He carried too many strings to his bow. He had professional knowledge and the gift of expression'.[20]

Next they took the government party across the Fremantle Bridge to the Richmond site, a drive to the east of Fremantle through the drizzling rain and along muddy tracks. 'There they made their last and most effective stand.'[21] The area was a green strip of country with bright fresh-looking winter grass, free from rock or timber, and was available for £30 per acre. Surely the Engineer-in-Chief would support the use of this site—especially as it had been previously surveyed by engineer R. W. Young, who had said that the workshops could be adapted to the locality. But O'Connor was not to be won over. When pressed, he agreed that the site might offer reasonable facilities but at greater expenditure than the Midland site. 'It is a principle', said Mr O'Connor

> in building workshops that you should get the shops to suit the land
> and not the land to suit the shops. I do not say that the land could
> not be made suitable but I would not agree to the new plans which
> have been made [by Mr Young].[22]

O'Connor added to this and other objections to the site by saying that every acre required for extensions in the future would mean heavy earthwork at great expense. As at Rocky Bay, the arguments for the Richmond site were countered by O'Connor's superior knowledge of the facts and the figures at his fingertips. The arguments were orderly, but the Fremantle-ites, led by W. E. Marmion, soon knew they were outgunned and decided that their best plan was to regroup to fight another battle later. When O'Connor proposed further investigation of the site from a higher vantage point, the majority, now dispirited, bedraggled and wet, voted for adjourning to the Fremantle Town Hall, where light refreshments had been prepared for the party.

> The general impression was that the Fremantle representatives had done their utmost in putting the case forward and that the materials for a good fight on both sides had been got together, of which more will be heard before the session [of parliament] is much older.[23]

The arguments dragged on, encouraged by the government's hesitancy. Although Venn had successfully introduced a motion to move the workshops from Fremantle to a site near Midland as early as September 1895—his motion being passed by seventeen votes to eleven—the work was not proceeded with. It took another six years and a Committee of Inquiry reporting in December 1901 before work was begun in earnest to make the transfer. It had been a bitter struggle but, as the *West Australian* had pointed out:

> It was so with the Fremantle Harbour Scheme, it was so with the Yilgarn Railway, it will be so with the removal of the shops...the Engineer-in-Chief has been successful in convincing the most unwilling of his opponents.[24]

However, O'Connor was never to see the Midland site truly operational.

Ironically, the Midland Junction Railway Workshops were to become a leading industrial and much respected enterprise in the colony, resulting in the development and increased population of the Swan Valley area west of the Darling Range. When the shops were closed finally, with the coming of electrification a hundred years later, the struggle to save them, together with the jobs of the employees, was long and bitter and the arguments used were much the same as those raised by the Fremantle-ites in the 1890s.

We may wonder how C. Y. O'Connor, the stern public man who worked unceasingly, controlling an expanding department, planning large public works and confronting stinging and often ill-informed criticism, could maintain a necessary equilibrium. There seemed to be no hint of depression at this stage of his life, yet the strain on him must have been at times overwhelming. The suggestion we have that he could fly into a rage is provided by his daughter, Kathleen O'Connor, who wrote, as noted earlier, in her memoir that he would dash his hat on the ground and dance on it. 'I never saw him do it myself', she added. But apparently others had. She then provides a clue to the nature of his restorative: his family life. She suggests that theirs was a happy household, that her father had a sense of humour and 'took much pleasure in his home'.[25] One of the more pleasurable family occasions—though perhaps tinged with parental sadness—and a welcome interruption to his professional duties, would surely have been the marriage of his eldest daughter, Aileen. Her groom was Charles Young Simpson, son of J. B. Simpson, a colliery magnate of Northumberland, England. The young Simpson, whose initials the family was amused to note were the same as their father's, travelled out to Western Australia on the same ship as the Governor, Sir Gerard Smith. Smith had been so impressed with Simpson that he had offered him a job as his social secretary. In the newspaper reports of the Governor's receptions and garden parties, where the presence of the O'Connors included Aileen, the name of C. Y. Simpson is also prominent.

Exactly why Charles Simpson had left England is not clear, although the existence in the archives of a gossipy letter from O'Connor's sister Frances, of Ballycastle, provides a possible explanation. A year before the wedding, and therefore probably when the engagement was announced, Frances had warned her brother that Charles Simpson's stepmother was 'very disagreeable and common' and that everyone was blaming her for 'turning the sons out of the house and the country'. She then wrote, 'With a woman like that with so much power I would be very careful how I trusted Aileen to them'.[26] She advised her brother not to allow the marriage. Although the correspondence between Frances and her brother in Australia was frequent and affectionate, she addressing him as 'Dear Charlie' and he writing often, sending her newspaper accounts of his work and photographs, on this occasion her advice was evidently ignored.

The occasion of the wedding in Fremantle was described as 'one of the most largely and fashionably attended weddings that have been witnessed in the colony'.[27] In balmy autumn weather on 1 May 1895, a crowd of some 150 family friends, professional associates of O'Connor, and members of the government packed the little Gothic-style church of St John, which had been completed only thirteen years previously. Heading a roll call of the most influential families in the colony were the Lieutenant Governor, Sir John and Lady Forrest, the Venns, the Shentons, the Lee-Steeres, Mr and Mrs Septimus Burt, the Drummonds, the Marmions and the Parkers. Afterwards the wedding party was received at the O'Connor home, where 'light refreshments were partaken of'. All the dresses of the women present are described minutely, and the presents given by each of the guests are also listed. Mr O'Connor gave a cheque of unspecified value. The bride and groom sailed soon afterwards for England to stay at the Simpson family home, 'Bradley Hall', in Northumberland. Whether Aunt Frances's fears for Aileen's reception there by the second Mrs Simpson were justified is a question that must remain unanswered.

18

THE CLOSING MONTHS of the year 1895 are significant in the story of C. Y. O'Connor because it was then, at the request of the government, that he first gave his full attention to a pipeline scheme to bring water to the goldfields. As noted in chapter 17, two separate proposals had been aired the previous year but had not been proceeded with. The first was contained in a letter to the *West Australian* of 8 March 1894 over the initials J. S. T. It suggested, in part, that water could be brought to the fields by means of a pipeline aided by pumping stations erected at Northam and along the railway line:

> Looking at the place with an unprofessional eye, there does not appear to be any physical difficulty that it would be impossible to overcome, and if competent authority, our Engineer-in-Chief for instance, were to pronounce the undertaking practical, there ought to be no difficulty in finding the pecuniary means to carry it out without appealing to the Government of the colony for aid.[1]

J. S. T. then argued that the construction could be paid for by selling the water at Coolgardie and outlying districts at a shilling a gallon.

The second proposal, similar to the first, was made five months later by a Mr John Maher, who applied to the government for the right to construct a pipeline from the vicinity of Northam and along the then incomplete railway line to Coolgardie. Maher and his associates were confident that they could raise the necessary funds—by their estimate, £2,500,000—on the London money market. They believed that they would be able to sell the water on the goldfields for 3s 6d per 1,000 gallons. Maher was told that in order to proceed, he would have to find someone to introduce into the Assembly a private member's Bill to authorize Maher's syndicate to do the work, or, if this failed, one to authorize the government to do the work or contract it out. Maher's efforts seemed doomed at the start because 'no member of the Assembly could be found to introduce either of the Bills because all regarded the project as a mad proposal, and would not be associated with it'.[2]

Maher, not to be discouraged, was persuaded to take his scheme to Alexander Forrest but was astute enough to withhold important details. Forrest showed interest, although he insisted on knowing the full plan. After this was disclosed, albeit reluctantly, by Maher, Alexander took the plans to his brother, the Premier. (So influential was Alexander and so close to the Premier that he was dubbed 'the sixth minister'.) According to one report, John Forrest was sceptical, dismissing the proposal as 'mad and impracticable...it must have emanated from the brain of a lunatic at Fremantle (i.e. in the asylum there)'.[3] According to the same source, John Forrest put the papers into his pocket and showed them later to the Engineer-in-Chief, who pronounced the plan 'perfectly practical, perfectly feasible'. It was later alleged that Sir John Forrest was guilty of appropriating Maher's idea for his own purposes and never once revealed whence he got his conception of the great water scheme.[4] Whether the accusation had any truth in it or not, Forrest from that time forward appears to have accepted a pipeline scheme of some kind as the only reliable method of bringing sufficient quantities of water to the goldfields. This was borne out by his hurried introduction of the Water and Electric Works Licence Bill into the Legislative Assembly on 8 October, only four days before the end of the parliamentary session. The purpose of the Bill, ill defined by its ambiguous title, was to smooth the way for

> any person or corporation to lay down water mains on or under any land, whether vested in the Crown or not...for the purpose of conducting water to be afterwards used and sold by the undertakers.[5]

In simple words, it was a Bill enabling either the government or private enterprise to build a water pipeline to the goldfields—a notion clearly in the Premier's mind. In arguing for the Bill, Forrest said:

> It is to be regretted that there is not a large quantity of fresh water available in that auriferous district, but the fact remains that there is a great scarcity, and that scarcity is not likely to be overcome, at any rate by boring. That being so, there seems to be no reason why we should not encourage persons who are willing to lay down water

mains from the rivers near the coast to that arid though very
auriferous district.[6]

In commenting on the Bill, the *West Australian*, after dismissing the likelihood of artesian water being found on the goldfields—no more likely than
'extracting the precious liquid from the quartz itself'—went on to endorse the
alternative suggestion, 'the most promising of all those yet submitted, [which
was] to carry some of the superabundant rainfall on the west side of the Darling
Range to the drought-stricken interior'.[7]

That Forrest favoured the latter course by this time is evident; the only
uncertainty in his mind was whether such an expensive scheme was best undertaken by private enterprise or by the government. Most likely he doubted
whether raising the enormous government loan that would be required would
be politically sustainable. The careful, judicious, logical Charles Harper, who led
an informal fledgling Opposition in the Legislative Assembly, proposed that
because the question was of such extreme importance and the ramifications so
enormous, the government should 'examine the feasibility of pumping water
from the most accessible spot, and prepare an estimate of cost and working
expenses with a view to its being undertaken by the State'.[8] Although this
proposal was outvoted in committee, it forced Forrest to guarantee that such
investigations would be undertaken. And it was this guarantee that would bring
C. Y. O'Connor back into the picture. Once again, Forrest turned to his
Engineer-in-Chief for practical support. Upon O'Connor's investigations and
recommendations would the future of a pipeline scheme be determined.

There is some evidence that O'Connor had started investigating the
feasibility of a pipeline before Forrest's guarantees were given in the Assembly,[9]
but certainly for a concentrated period of two months from the middle of
October 1895, and continuing at a lesser pace until the reopening of parliament
the following July, the officers of the Public Works, led by their Chief, worked
tirelessly on collating technical information, statistics and estimates of costs,
drawing up plans, and investigating the manufacture and quality of pipes
and pumping machinery. Surely no enterprise—at least until that date—had
ever received more searching investigation than this one. O'Connor was the
captain, he issued instructions, and all information was submitted to him
for scrutiny.

The first file, comprising papers relating to the selection of the pipes that
would carry the water the 350 miles, includes evidence from around the world
of the relative merits of wrought iron, cast iron and steel; and tables of comparative costs of pumping 5 million, 10 million and 100 million gallons of water per

day.[10] In each case, the data are assembled by O'Connor's assistants, Arthur Dillon Bell, Horace Robertson and others, and often their work is returned for clarification. 'I do not see how I can compare these sets of prices as they are apparently on different bases as regards the pressure on the pipes', he writes on one document; and on another, 'Please see me and bring with you tabulated statements—I want the whole lot now put in type with latest revisions'. He investigates the tensile strength of steel, which he finds is one and a third times stronger than wrought iron; steel was widely used in the United States. He abhors blueprints 'chiefly for the reason that it is almost impossible to make corrections on them'. The files dealing with pumping equipment, estimates of costs, specifications, and consumption of coal are probably the most fascinating, involving cables sent and received in code from overseas.[11] He is concerned about the poor quality of Collie coal—one and a half times more was required to do the same work as Newcastle coal, on which certain calculations had been based—and the costs of shipping the pumps and assembling them in Western Australia. Every detail is thoroughly investigated, and comparative sets of figures are prepared to cover a range of options: whether the government will eventually choose this or that quantity of water; this or that route for laying the pipes; this or that number of reservoirs feeding water to towns along the route. O'Connor was conscious that he was not required to make policy decisions; his duty as a public servant was to furnish the government with accurate information so that the right choices could be made in parliament.[12]

O'Connor, although often required to travel during this period, and frequently immersed in his other projects, kept himself fully informed, and continued to issue instructions to his staff and make his own summaries even while on the move. The enormity, thoroughness and accuracy of this investigation are one of the lasting achievements of C. Y. O'Connor—ably assisted by his staff, it must be said. Somehow, at the same time, he also managed to deal with railway and harbour business, including the investigation of a serious fire in Fremantle that destroyed one of the goods sheds. It is little wonder that the *West Australian* of 13 November 1895 reported that the Engineer-in-Chief had been suffering from ill-health brought on by overwork. No symptoms of his illness are stated, and in the absence of more information we are led to wonder whether this was a harbinger of the breakdown he was to suffer from overwork and stress six years later, which led to his untimely death.

However, on this occasion his 'ill-health' was evidently not of lasting seriousness because he prudently combined a rest cure with work by escaping from the cares of his city office. He sailed on the *Sultan* for a visit to Carnarvon, 'where he will inspect the harbour with a view to advising the Government on improvements'.[13] But even while in Carnarvon, he was still 'the Chief', receiving reports on the Goldfields Pipeline Scheme and cabling his staff for clarification of certain details. He returned to Perth via Geraldton three weeks later, having advised Carnarvon on a relocated and more efficient shipping facility resulting in a substantial reduction in the estimated cost of building a stock jetty.

Whatever might have been the nature of O'Connor's ill-health, as minimally reported in the newspaper, it must have acted as a warning, convincing him that the addition of the pipeline project to all his other works required him to shed some of his load. There is little evidence that O'Connor favoured the pipeline scheme solely because it exalted his reputation, as some of his critics have argued, but it would have been perfectly natural for him as an engineer to have been inspired and excited by a work that was considered one of the engineering marvels of the Empire at that time. It is of little surprise, therefore, that at the conclusion of his report to parliament on the working of government railways for the years 1892–95, dated 13 August 1895, we find a plea that his burden be lightened. He reminded the government that while the decision for him to act as general manager of the railways was reasonable at the time he was appointed, 'the time has arrived or is, at any rate, near at hand, at which I might be relieved of this responsibility'. In defence of his position, and with a touch of humour unusual in a government report, he uses some of those very arguments that had been raised against him by disgruntled contractors and officials:

> Concentration of a very wide range of work in the hands of one officer necessarily involves delay and inconvenience to the persons who have a right to see him, and I am painfully conscious of the fact that I have had to inconvenience many people recently in this way. They have been, so far, very amiable under the infliction, but even suffering humanity is apt to rebel at last.
>
> In view, therefore, of the large and rapidly increasing volume of construction works, to which my attention is more immediately due, it would be a great relief if I was permitted to be entirely free from any responsibility as regards the organisation and working of the opened railways.[14]

It is worth noting that in this last sentence, he refers only to the management of 'working' railways. He did not envisage relinquishing his responsibility for planning and constructing new railways.

The *West Australian*, after chiding his department for the long gap since the previous report on the railways, wrote glowingly that 'it is something far

short of sufficient praise' to judge this report 'the best thing of its kind which
has appeared in connection with the West Australian railway system'.[15] Among
the improvements that O'Connor had been able to introduce, and which the
newspaper so strongly applauded, were: the replacement of narrow light rails
with 60-pound tracks allowing heavier and safer rolling-stock; the increase of
total railway mileage, which had been 183 prior to 1890, to 573; the completion of the Yilgarn Railway as far as Coolgardie; the commissioning of the
South-Western line to Bunbury, Boyanup and Busselton; and commencement
of the 20-mile Mahogany Creek Deviation, reducing the ruling grade from 1 in
15 to 1 in 50. From the loss-making railway system that O'Connor inherited,
he was able to show that by 1895 the net revenue amounted to £113,954 and
the net earnings per train mile had increased from 0.3 pence per mile in 1891
to 27.43 pence per mile in 1894–95. He attributed this dramatic improvement
to the heavier rails and heavier locomotives.

Although he was able to show that improvements had been made in the
railway workshops, he repeated his argument for their removal to a larger and
more convenient site:

> The [present] workshop is practically filled up with machines, leaving
> hardly any space for vehicles which have to be repaired, the result
> being that nearly all the repairing work has, for want of additional
> shed room, to be done in the open.[16]

The report goes on to list the many improvements made to station buildings,
signalling and water supplies throughout the system; the increase in train
services; and the improved grades on the Eastern line. At one point in the report,
O'Connor expresses his concern to improve the education of cadets in the service
and says that it would be desirable for them to be provided with university
training 'as in all the adjoining colonies'. (O'Connor's eldest son, Frank, was at
that time a cadet in the department.) He mourns the death of his 'old and valued
friend' F. W. Martin, who was engineer-in-charge of the working railways;
another loss deeply regretted is the 'valued officer' F. Calver. The newspaper
commended O'Connor for acknowledging credit due to his staff, particularly the
traffic manager, John Davies. The *West Australian* thought that in the management of the railways, there were still many weak spots and that 'a most watchful
scrutiny will always be necessary, but for honest, capable and efficient work a
high record is due to the officers of the West Australian railway department'.[17]

Forrest's original cable confirming O'Connor's duties in Western Australia
with the careless-sounding words 'Railways, Harbours, everything' must have

echoed somewhat ironically in the Chief's memory as he wrestled with his many responsibilities. Much of his routine engineering work has been overshadowed by his major works; some is nearly forgotten and difficult to trace, such as his work on Rottnest Island's main lighthouse on Wadjemup Hill. We find the Chief and his wife present at the ceremonial laying of the foundation stone on 25 April 1895. They had travelled as members of the official party on board the SS *Dolphin*, 'a pleasant run of one hour twenty-five minutes'[18]—but a slow crossing from Fremantle by today's standards. O'Connor was the supervising engineer, and his earlier marine experience in New Zealand would have made him a formidable expert guiding his assistant, George Temple-Poole, and the builders, Parker and Rhodes. The lighthouse was completed and opened a year later but the Chief was not present at the celebrations on that occasion.

In Fremantle, the harbour work was proceeding but had been hampered during the winter months, according to one report, by 'exceedingly rough and continuous wet weather'.[19] Blasting operations were being carried out so success-fully that much of the solid rock blocking the entrance to the river had been cut through by 30 June 1895. A second bucket dredger, the *Parmelia*, and a second suction dredger, the *Governor*, were added to the fleet of four in 1896. The original dredger, the *Fremantle*, was joined by the *Premier*, from Albany, and together they moved through the bar and started working on the sandy bottom of the inner basin.[20] Sir John Forrest inspected the work on 14 November 1895 and expressed himself as being highly pleased with the progress that was being made.[21]

In the early months of 1896, leading up to the opening of parliament in July and beyond, O'Connor continued exhaustive research for the anticipated start of the Goldfields Pipeline Scheme. A key member of staff was hydraulics engineer Thomas Cowley Hodgson, who had arrived in the colony from Melbourne in March the previous year and was appointed engineer-in-charge of sewerage and water supply within O'Connor's overall control. Hodgson proved himself an able and trusted colleague, giving accurate and much welcome advice to his Chief. Hodgson's office was in the old Bon Marché Arcade in the centre of Perth, and the exchange of memoranda in the files between Hodgson and O'Connor in his Treasury Building office shows how much O'Connor relied on Hodgson and respected his judgment. As yet there was no hint of Hodgson's eventual fall from grace, which would prove such a decisive factor in O'Connor's destiny.

Another factor that could be seen in retrospect as having a small part to play in O'Connor's destiny was the dismissal of Harry Whittall Venn from the ministry. The Engineer-in-Chief had always kept himself aloof from parliamentary squabbling and political gamesmanship, preferring to see his role simply as that of a loyal public servant. But he must have watched with growing concern the deteriorating relationship between the Premier and the Minister for Railways and Public Works in the early months of 1896. For five years, a close working partnership had developed between Venn and O'Connor based on mutual respect and shared objectives, the former a strong ally of O'Connor's in Cabinet and in the Assembly. However, Cabinet reshuffles to save political reputations were as common then as they are today. Venn was on record as saying that there was never any dissension in Forrest's ministry and that the longer he and his colleagues were associated with the Premier, 'the better we appreciate him and the more loyal we are to him'.[22] The irony of that fulsome declaration must have haunted Venn in the early months of 1896.

The root of Venn's quarrel with Forrest was the growing shortage of rolling-stock in working order, which was recognized in 1895 as seriously impeding the progress of the railways. Demand had been increasing for trucks to carry water and equipment to the goldfields; thousands of tons of cargo awaited transport from Fremantle. 'The whole port was in chaos.'[23] The crisis had been aggravated in October the previous year when the fire in Fremantle had reduced the already inadequate supply of trucks by twenty-four. Both Venn and O'Connor, supported by Davies, the traffic manager, had pressed for further expenditure on rolling-stock, but Forrest had delayed authorizing necessary funds for this purpose. So great was the number of citizens—merchants, prospectors, construction workers and passengers alike—who were affected by the crisis that a public meeting was held in the town hall. Blame was firmly laid at Venn's door and the delegation demanded that Forrest replace him. Forrest was non-committal in front of the delegation but later defended the government by publishing correspondence suggesting that Venn was to blame for the shortage of rolling-stock, that his officers 'were afraid of overdoing it by ordering too much'.[24] Venn was stung by the attack on his competency and unwisely retaliated by stating his case in the press, calling into question Forrest's honesty.[25] The dissension between the minister and the Premier became a public scandal and led immediately to the Premier's call for Venn's resignation three times by letter and once by telegram. But Venn remained obdurate. Forrest then had little alternative than to ask the Governor to dismiss Venn. He did so by telegram, which was said to have arrived after Venn had gone to bed. Later a bitter Venn claimed that 'he had been dismissed in his nightshirt'.[26]

O'Connor's new minister was Fred H. Piesse, a miller and merchant from Katanning. Piesse was a steady, methodical man, not brilliant but showing himself capable of mastering the requirements of this most difficult of portfolios.

(When Venn was first appointed, he said humorously that 'he looked forward to being the best abused man in Western Australia'.)[27] Piesse's opposing wags in parliament, familiar with the Book of Common Prayer, dubbed him 'the Piesse which passeth all understanding'.[28] There is no record of what O'Connor thought about the change in the ministry, but there is little doubt that he lost in Venn a faithful if sometimes stubborn and tactless ally who would have given him valuable support in the crucial years ahead.

Throughout 1896, O'Connor continued assembling data in support of the pipeline, and much of the correspondence to and from the minister, once signed by Venn, now bore the signature of Piesse.

O'Connor's old and valued friendship with one-time head of the Department of Works in New Zealand, John Carruthers, was renewed by correspondence in March. Carruthers, now Sir John, had since been appointed Consulting Engineer in London to both the New Zealand and Western Australian Governments, and O'Connor placed before him the pipeline scheme, asking for his advice. Carruthers's reply of 24 April comments favourably on the plan, suggesting which manufacturers should be approached for pumping equipment and agreeing with O'Connor that the pipeline should follow the track of the railway. If not, he writes

> Just think what an array of mules and jiggers you would require to cart the pipes. The section of the railway is not very favorable, nor is it very unfavorable, so that you will have no special difficulties in following it.[29]

Before the opening of parliament in July, O'Connor clearly wanted to present a case for the pipeline that was unassailable from a practical engineering perspective. To this end, he submitted his data to a number of authorities and received their endorsement. Within his own department, Hodgson, a recognized authority on hydraulics, wrote:

> I consider the whole scheme much simpler than the one I carried out in Tasmania a few years ago. There we pumped 420ft high in one lift, and the pumps were worked by turbines driven by the water of a river and without steam power.[30]

The commitment of Forrest's government to the pipeline scheme was first made public in the Governor's Speech at the opening of the third session of the Second Parliament on 8 July 1896. After making reference to the planned Constitution Convention and the colony's participation, the Governor, Sir

Gerard Smith, turned to the matter of the goldfields and the necessity for a permanent water supply:

> A Bill will be submitted to you for the purpose of providing loan funds to enable five millions of gallons of water to be pumped daily from reservoirs to be constructed in the Darling Ranges, and taken along the railway line to a reservoir on the summit of Mount Burgess and there distributed by gravitation to Coolgardie and Kalgoorlie, and other places.[31]

Nine days later, Sir John Forrest introduced the Coolgardie Goldfields Water Supply Bill to raise £2,500,000 for the construction of the pipeline, and on 21 July, armed with a comprehensive report from his Engineer-in-Chief, he spoke for two hours, moving the second reading of the Bill. Quoting the prophet Isaiah—'They made a way in the wilderness and rivers in the desert'— Forrest said that future generations would praise his government and the parliament for such a far-seeing enterprise. He outlined the details: building a weir on the Helena River in the Darling Range, and pumping water 350 miles overland to Mt Burgess. Eight pumping stations would deliver 5 million gallons a day at an estimated price of 3s 6d per 1,000 gallons.[32] The Premier pointed out that the scheme would not only supply the mines with industrial water and the residents with drinking water but would serve the railway system and possibly the towns of York, Northam and Southern Cross. He also stated that investment in the mines and in agriculture in the areas served by the pipeline would increase, and that the confidence of overseas investors would be restored when they observed that the mining industry was to be provided with an adequate supply of water. Forrest concluded that he would rather go out of office than give up the scheme.

The second reading debate carried over into a second week, but with the display of such irresistible confidence on the part of the Premier, backed by the technical data supplied by O'Connor, the critics in parliament had a hard struggle to raise valid objections. What objections there were centred mainly on the ill-founded belief that water already existed in sufficient quantities below the surface, or in lakes at Broad Arrow; other concerns were that insufficient water could be pumped from the Darling Range for the entire needs of the mining areas, or that the huge cost of the project made it simply uneconomical. Support for the measure was so solid, however, that the second reading was passed in the Legislative Assembly on 5 August by a show of hands, a division being found unnecessary.

The Bill's progress through the Legislative Council was equally if not more comfortable. The second and third readings passed on the same day,

3 September. The ease of the Bill's passage contrasted oddly with later criticism as the scheme progressed.

O'Connor's work in amassing technical information did not cease with his report to the Premier in July. By this time, he knew well the ways of government, how at first there were long periods of delay and indecision, and then, once a Bill had been passed, there were demands for instant action. He had written seeking John Carruthers's advice in March, urging a quick response because 'If the Government were to adopt the scheme...as is usual in such cases [they would] expect it to be put in hand with lightning rapidity'.[33] He had also warned his staff that Forrest sometimes wanted to do things in 'a mighty hurry scurry'.

As if to prove O'Connor's point, in early November Forrest sent a handwritten note to Piesse reminding him that 'it is now three months since the Water Supply Bill was approved by Parliament...I should like to know the present state of business and what progress has been made'. He said he had hoped by this time that orders would have been made for pipes and pumping machinery. He reminded O'Connor, through his minister, that 'the main reservoirs should be taken in hand as soon as possible'.[34] We also observe in this note the first reference to O'Connor being sent on a trip to Europe in connection with the scheme. 'It seems absolutely necessary that the Engineer-in-Chief should go to London at once', Forrest wrote, and he wondered whether the delay in departure had adversely affected progress of the scheme. He concluded his note affably enough by assuring Piesse (O'Connor) that he was not really complaining of any loss of time, but the great responsibility resting upon him and his anxiety in regard to the great project 'makes me press the matter upon you'.[35]

O'Connor answered Forrest on 14 November, reminding him politely that it was not three months since the Bill was passed in parliament but only two (Forrest had calculated from the Bill's passage in the Assembly; O'Connor from its passage in the Council). He summarized the work completed: the survey of alternative routes; the compilation of the longitudinal section along the proposed line of pipe; the detailed surveys of the reservoir at Mundaring; working drawings of the reservoir; and the contour surveys for the terminal reservoir at Mt Burgess. He says that there had been delays in the lithographic department, their work so necessary for the printing and circulation of plans, but he hoped that the work would be completed by the end of November. He assures the Premier that the delay in departing for England had not adversely affected progress, but to the contrary: he was forwarding the plans for the pipeline to a Commission of Engineers in London for their opinion and 'If I happen to reach England contemporaneously with the said drawings I would join in the Conference and wire the Government my advice'.[36]

O'Connor was able to send off his report and several books of lithographed drawings to John Carruthers in London for forwarding to the Commission of Engineers on 4 December. The final estimate he gave for the project was then £2,834,000.

On 11 December 1896, he sent off additional notes on the scheme to Carruthers. By the end of the month, Carruthers wired the names of the eminent engineers who would sit in judgment over the pipeline scheme and asked for authority to pay them 50 guineas each plus an additional 12 guineas per diem allowance. O'Connor cabled back the same day, expressing his satisfaction with the selection of the committee members: Professor William Unwin, 'probably the highest authority of the day on the transmission of power', and Dr George F. Deacon, engineer-in-charge of the Liverpool Water Supply—'one of the largest things of its kind that exists'.[37] The third member of the commission would be John Carruthers himself.

And so a momentous year in the history of the pipeline and the life of O'Connor comes to a close. The surviving files, bulging with correspondence amassed throughout the year on the pipeline scheme alone, are graphic testimony to the mountain of work that O'Connor and his staff accomplished. By December, the first stage—the research—had been completed. The second stage—the construction itself—would occupy the next five years, not three as O'Connor had confidently predicted in his report, although the blame can hardly be laid at his door.

For much of 1897, the Chief would be away overseas but he would leave his department in experienced hands: A. W. Dillon Bell, acting Engineer-in-Chief; T. C. Hodgson, engineer-in-charge of construction of the pipeline scheme; and William C. Reynoldson, assistant engineer. In the ensuing months, and even continuing into the next century, questions would be raised and debated about the origin of the scheme. Who, it would be asked, might claim the honour of originating the idea, and who was the most persuasive advocate of that scheme? Even today legends abound and various misconceptions endure.

Surely no uncertainty need persist. In their definitive and meticulously researched inquiry into the origins of the scheme, Professors F. Alexander, F. Crowley and J. D. Legge conclude by stating that the Goldfields Pipeline Scheme 'may well be described as politically Forrest's, and technically O'Connor's though neither man was responsible for the original idea', and in answering the question as to who was responsible for placing before Sir John Forrest the plan that was adopted, the answer, firmly given, is 'C. Y. O'Connor and his departmental officers'.[38]

19

Before we follow O'Connor on his mission overseas, it is timely to take a closer look at the Goldfields Pipeline Scheme that he devised, and consider not only the criticisms that were made at the time but the more recent and more serious charge that he knowingly withheld information from his report so as to strengthen the case for a project that would greatly enhance his professional standing.[1]

Conveying water over long distances was not new at that time. The earliest known aqueducts were built by the Romans—one carrying water a distance of 25 miles mostly through tunnels and most famously across the three-tiered Pont du Gard near Nîmes in Southern France. In the second half of the nineteenth century, several bold schemes were in place or in the course of construction in Britain. The most notable was the 68-mile pipeline, commenced in 1881, drawing water from the dammed Vyrnwy Valley in Wales to supply water to the Birmingham district; another, in 1885, was a pipeline from Thirlmere in the Lake District, supplying water for Manchester 95 miles further south. O'Connor must have been familiar with the construction details of both these projects.

What was unique about O'Connor's pipeline at that time was the great distance that the water would travel, and the requirement to pump the water *uphill*. The approach to the goldfields was a long and gentle but substantial rise. Coolgardie is 1,400 feet above sea level, and the water at Mundaring is 340 feet above sea level, which gives a total net lift of over 1,000 feet. O'Connor calculated that the total gross lift *including friction* was 2,632 feet. To pump 5 million gallons of water per day a distance of 329 miles at the required pressure, eight pumping stations were planned, the first close to the main storage reservoir, which has been preserved and can still be seen in the C. Y. O'Connor

Museum at Mundaring. Originally, there were three sets of boilers, engines and pumps—two working sets and one spare. The second pumping station was to be located at about 1½ miles from the first, with machinery similar to that at number one, capable of raising the water to a high point in the Darling Range from which it would gravitate a distance of 77 miles to pumping station number three (Cunderdin). The fourth pumping station would be located at Merredin, 64 miles from number three. Stations five to eight were to follow the railway line: number five at Yerbillon; number six at Ghooli, 8 miles east of Southern Cross; number seven at Gilgai; and number eight at Dedari, which would pump water the final leg to the Coolgardie reservoir, a total distance from Mundaring Weir of 329 miles. The original plan did not include the 24-mile extension to Kalgoorlie. The total lift at each of these pumping stations was approximately 225 feet. Since Kalgoorlie was lower than Coolgardie, the water on its last stage of the journey would be fed by gravitation. O'Connor had worked out that to supply 1 million gallons daily would cost £1,000,000 (taking into account smaller pipes and less powerful pumps) but to supply 5 million gallons would cost £2,500,000—that is, five times the quantity of water for two and a half times the cost—and that the cost of the water to the consumer for the smaller scheme would work out to 5s 6d per 1,000 gallons compared with 3s 6d for the larger scheme.

O'Connor set out his scheme in the minutest detail in a twenty-three page document entitled *Report on Proposed Water Supply (by Pumping) from Reservoirs in the Greenmount Ranges*, and it was this report of which Forrest made such thorough use in his second reading speech to parliament. He quoted extensively from it verbatim; for this reason, much of the criticism of the scheme that was to follow was directed at O'Connor himself. O'Connor had anticipated that this might be the case in writing his report. In his opening paragraph, he refers to the misconception that it was a proposal that he was urging on the government:

> I need scarcely say that I never urged, nor do I now propose to urge
> upon the Government or the country, the undertaking of this work,
> but while it would evidently be quite improper for me to do that, it

OVERLEAF: Sectional plan (longitudinal profile) of the Goldfields Pipeline showing the positions of the pumping stations and other details. The rise and fall of the ground over which the water was to travel is clearly shown in the cross-sectional drawing. The rivetless or locking-bar method of constructing the cylindrical pipes is illustrated top left.

Original held at SROWA

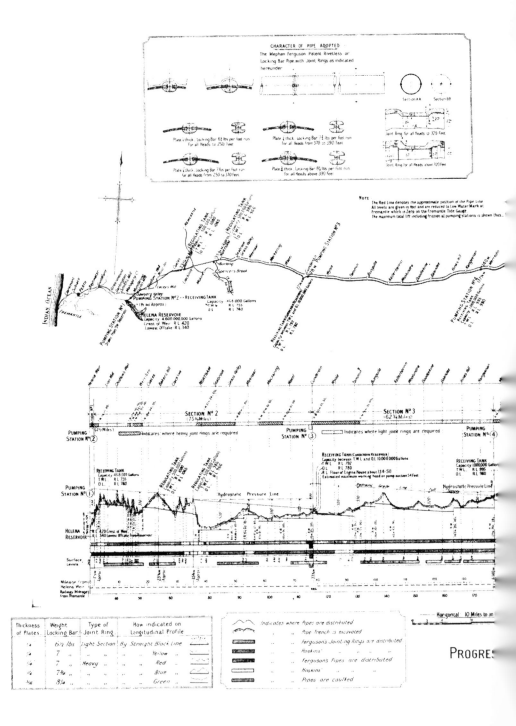

OLGARDIE WATER SUPPLY

N AND LONGITUDINAL PROFILE OF THE PIPE LINE.

ewing location of & pressures on different classes of
es, excavation of trench, distribution of pipes & rings,
& caulking operations &c.

— PLAN —

EXPLANATION
R L denotes reduced level
T W L top water level
O L offtake level
Mileage from Helena Reservoir along the pipe line is given in red figures
Mileage from Fremantle along the railway is given in black figures

PUMPING STATION N°6
Receiving 1,000,000 Gallons
O L R L 1325

PUMPING STATION N°7
Receiving 1,000,000 Gallons
O L R L 1411

PUMPING STATION N°8
Receiving 1,000,000 Gallons
O L R L 1447

—UDINAL PROFILE—

| SECTION N°5 (46 Miles) | SECTION N°6 (51¾ Miles) | SECTION N°7 (45 Miles) | SECTION N°8 (32¾ Miles) |

PUMPING STATION N°6
RECEIVING TANK
Capacity 1,000,000 Gallons
T W L R L 1345
O L R L 1325

PUMPING STATION N°7
RECEIVING TANK
Capacity 1,000,000 Gallons
T W L R L 1431
O L R L 1411

PUMPING STATION N°8
RECEIVING TANK
Capacity 1,000,000 Gallons
T W L R L 1467
O L R L 1447

Hydraulic Grade Line pumping ½ Supply 5 to 7

Hydraulic Grade Line pumping for supply 6 to 8

Vertical 500 Feet to an Inch

Copy of P.W.D.W.A. **9037**

to 31.12.1901

Shews that there are sufficient of ferguson's pipes in Depôts to cover distance shewn on profile
.... Hoskins'
.... ferguson's and rings
.... Hoskins

is equally evident that if I am called upon to give an opinion as to the best way of attaining a certain object, it is clearly my duty to give such opinion to the best of my ability.[2]

In the third paragraph, he states that the proposal to pump 5 million gallons of water to the goldfields was frequently referred to as his proposal, but, he argues, it could only properly be held to be so in so far as it being in his opinion the best means of attaining the object in view and 'not in any way as being a fancy project of mine (as some persons have described it) which I am desirous of thrusting upon the Government and the country'.

The main criticism of the scheme, although insufficient to delay the passage of the Bill through parliament, was directed at its great cost. In the Legislative Assembly, George Simpson, Member for Geraldton, argued that the scheme was 'absolutely unnecessary', and that there were ample supplies of water beneath the ground.[3] He also alluded to the Premier's perceived desire to bolster his electoral support in the goldfields. Although there was no formal Opposition party at the time, S. H. Parker, in the Legislative Council, rallied opposition to the scheme by arguing that it was an unnecessary burden on the taxpayer, it would not be finished in the time estimated, it was a fad of the Engineer-in-Chief's, and that the report 'was not worth the paper it was written on'.[4] Generally, however, criticism at this early stage was fairly muted. The *Kalgoorlie Miner*, edited by John Waters Kirwan, was strongly in favour of the scheme. In its report of a mass meeting in the town held while the Bill was before parliament, 'Not a single voice was raised against the Government project from the vast concourse that assembled'.[5] Its neighbour, the *Coolgardie Miner*, whose editor F. B. Vosper later vociferously attacked O'Connor, thought that by the time the water reached the town, it would no longer be required. The *Northam Advertiser* called it 'O'Connor's quasi-amateur scheme', and the *Daily News* described it as 'a scheme of madness'. The *Morning Herald* of 30 October the following year maintained its continuing opposition to the scheme, believing it unjustified 'because we cannot afford it'. The *West Australian*, however, continued its support throughout.

As the months slipped by and the inevitable delays in the project arose—delays not of O'Connor's making—the criticisms in the press and in parliament directed at the Engineer-in-Chief would multiply to such a level that if they were not directly responsible for his death, they certainly contributed towards it.

There has always survived a minority view that the Goldfields Pipeline Scheme, in retrospect, may have been unnecessary and an overly expensive option. But the most serious charge against O'Connor appeared in 1993 with the publication of *The Golden Mile* by historian Geoffrey Blainey. Professor Blainey argued that O'Connor's brilliant scheme, when examined closely, 'was not necessarily an object lesson in the massive use of public funds'. He goes on to state that O'Connor 'seems to have mishandled the evidence'—namely, the rainfall figures for Coolgardie and Southern Cross as given in his report. These, he argues, were wrong, thus giving a false impression and exaggerating the need for the scheme to his own advantage. And he says that O'Connor 'probably knew he was wrong'. The conclusion is difficult to avoid, he says, '[that] O'Connor seems to have fiddled with the evidence'. Blainey then compares the prevailing water shortage in the Coolgardie goldfields with the similar situation and climatic conditions at Broken Hill, the rival mining centre in western New South Wales. He states that a Melbourne company built an inexpensive dam there in 1892 and that O'Connor must have known about it, although he did not mention it in his report. 'Had he mentioned it, he would thereby have weakened—but not destroyed—his own argument.'[6]

The case as presented by Professor Blainey is a fairly damning one; if justified it not only undermines the professional reputation of O'Connor but calls into question his honesty, to which his admirers, including Premier John Forrest—and even his leading critic, Alexander Forrest—had so often paid tribute.

To avoid what Blainey has accused others of doing—'sanctifying O'Connor' or 'treating him too sympathetically'—let us examine the evidence that is available, as it might be presented in court, conscious that we cannot be sure that the evidence available to us today was as clearly available to O'Connor in 1896.

The case for the prosecution rests mainly on two points. The first is O'Connor's use of rainfall figures for 1895, the year preceding his report, which gave a false picture of the water shortage. According to the charge, had the figures that he cited ranged more widely, they would have shown a more optimistic picture; by implication, a much cheaper scheme could have been devised—or a scheme might not have been necessary. The second point is the assertion that O'Connor chose to ignore the comparable water situation in Broken Hill and the more economical methods that the authorities had adopted there to solve the problem.

O'Connor began his report stating the case for pumping water from the 'Greenmount Ranges' to the Coolgardie district by comparing the prospects of rain in the two places. He states that he has no hesitation in saying that a large and permanent supply of water could be obtained 'in the cheapest and most certain manner' by pumping from reservoirs in the ranges, and then—quoting the rainfall figures—he gives the reasons that have led him to that conclusion.

He compares the average rainfall recorded in Perth during the previous ten-year period, 33.63 inches, with the known rainfall in Coolgardie, 6.79 inches, for one year—1895—which he states 'is the only complete year of which we have record and which is believed to have been an exceptionally wet year'.

The case against O'Connor at this point seems to be at its most damning. As Blainey points out, 1895 was not the only year for which rainfall figures were available. 'Surely a dutiful government official would have recorded [in his report] the rain at Coolgardie in several earlier years', he writes.[7] But O'Connor did not do so, and an alert prosecution counsel would point out to the jury that O'Connor remained silent about them because 'those records did not sufficiently help O'Connor's case'.[8] O'Connor also quotes in his report the rainfall for Southern Cross, which, he states, 'during the last two years has only been an average 5.26in'. Prosecution counsel would then ask O'Connor why he did not mention the 14 inches of rain at Southern Cross in 1893, and the 10 inches of rain in Coolgardie in the same year.

If O'Connor had an answer to that question, we do not know it. Did he merely accept the figures given him by a careless member of his staff? It seems unlikely that on this basic point he would not have checked them himself.

Prosecution could well have persisted, fixing O'Connor with an accusing look and paraphrasing Blainey's words: 'Should you not have made it clear that even on existing records Coolgardie was not as dry as you fiercely maintained?'.

In 1896, the year in which O'Connor was writing, the rainfall was already promising, and by the end of the season it reached 10.5 inches at Southern Cross and 8.5 inches at Coolgardie. Surely this would have been sufficient to fill dams, which could have been built at a far lower cost than a pipeline?

We turn then to the question of the comparison with Broken Hill, which, Blainey says, received just as much or as little water as Kalgoorlie: '[the town] eventually provided its own fresh supply without the aid of a pipeline'.[9] Blainey believed that O'Connor knew about the inexpensive dam built there in 1892 but did not mention it in his report because to do so would have weakened his argument 'that the heat and evaporation around Coolgardie were strong reasons restraining the building of a local reservoir'.[10] After pointing out that the small existing dams on the goldfields had been built in dry years and that therefore it was not quite fair of O'Connor to say that 'they had never been filled', Blainey concludes for the prosecution by writing that 'the evidence did not support a pipeline as large and as expensive as the one he planned, and defended literally to the point of death'.[11]

Counsel for the defence would surely open by reminding the jury that although rainfall figures for Southern Cross had existed since 1891, rainfall figures for Coolgardie had only existed for three years: 1893 (10 inches), 1894 (3.5 inches) and 1895 (6.79 inches). The average rainfall over those three years therefore was 6.76 inches—within a fraction of the figure that O'Connor cites for 1895. Even if the rainfall for 1896—which O'Connor could not have known at the time of writing—had been included, the average annual rainfall increases only to 7 inches, which is only a fraction more than O'Connor's figure. Significantly, this average is maintained over a seven-year period, 1893 to 1899. Although we cannot explain why O'Connor wrote that only one year's figures were available, we can say with confidence that he was, nevertheless, proved right about the average rainfall, and in the long term we can see how wise he was not to have been influenced by the infrequent higher than average figure. Even if he had included the rainfall figures for 1893 and 1894, his conclusion would have been the same and his argument would still be intact.

In writing of the dams that had already been built on the goldfields, O'Connor's argument, which Blainey does not quote in full, seems to be, at the very least, reasonable:

> We have selected the very best sites available, and although the reservoirs, in most cases, have only a capacity of one million gallons, there are many of them that have never been filled, and some of them into which hardly any water has come at all; and some of them leak very badly (in consequence of the kaolin formation which generally prevails in the district, having fissures in it through which the water escapes); and they have, moreover, been very costly, by reason of the cost of getting materials to them.[12]

The defence counsel might then call as witness the experienced hydraulics and sewerage engineer T. C. Hodgson, and ask him whether an average rainfall of 7 inches at Coolgardie, *an erratic and uncertain rainfall*, would be sufficient under the then existing ground conditions to keep dams filled with clean water all year round, given the need to supply a growing mining industry and increasing population. Hodgson would undoubtedly have answered 'no'. And this view is endorsed by modern engineers, who suggest that any area with a rainfall of under 10 inches per year would have serious water shortages and require similarly elaborate solutions.[13]

O'Connor's defence counsel might then produce one of the Chief's letters, dated 31 July 1894, addressed to engineer Mr Frank Reed, who was surveying and reporting on the location and the capacity of water tanks along the railway line. The letter shows conclusively that O'Connor was well aware that an exceptionally good rainfall in that area could be a deceptive indicator of a constant water supply.

> I quite admit, of course, that the amount of rainfall, which would be lost by evaporation and percolation is a very variable quantity, depending very largely on the character of the rainfall, but never-theless, some approximation of the truth can no doubt be made... I would be inclined to assume an average set of conditions rather than assuming either that there was a tremendous rainfall in a very short time, or that there was a very slight mist continuing for several days.[14]

Counsel for the defence might then turn to the question of the comparison with Broken Hill. Why did O'Connor not consider the parallel case of Broken Hill's less expensive alternative? Did he keep quiet about Broken Hill because the comparison might have provided the Coolgardie goldfields with a cheaper, less dramatic solution to its water shortage? But a good lawyer would have no difficulty in demonstrating that Blainey's argument was at its weakest here. The mean rainfall at Broken Hill over a ninety-one year period was 9.75 inches—significantly higher than at Coolgardie. The defence would likely have dismissed Blainey's assertions at this point and argued that had O'Connor cited Broken Hill as an example, far from weakening his case for a pipeline it would actually have strengthened it. Why? Because the early history of Broken Hill tells of water shortages and resultant hardships very similar to those recorded on the Coolgardie–Kalgoorlie goldfields. In the late 1880s, the local dam water was stated to be unfit for human consumption and 'the metallic dust on household roofs from smelter fumes found its way into the rainwater tanks'.[15] A scheme had been advocated for the construction of a pipeline from the Darling River as early as 1888, but the project was shelved owing to lack of money. Like Coolgardie, most of the water was then obtained from condensers. Water had also been transported at one time by rail, at great expense, from Mingary in South Australia. A dam was completed at Stephens Creek in 1892, 'but within a few years could not meet the demands of mines and the township'.[16] Water shortages continued to be a serious problem and various schemes were always being considered, but none of them succeeded in solving Broken Hill's acute water problems. A second reservoir was constructed in 1914.

> [A]lthough the combined capacity of both reservoirs was over 38,000 megalitres (84 million gallons) the problem of a constant source of

water still remained, *for in years of low rainfall, either or both reservoirs became dry.*[17] [emphasis mine]

Ironically, it was not until 1946, after numerous public meetings and pressure being exerted on the government, that the construction of a pipeline was commenced. When completed in 1952, it pumped water from Menindee to Stephens Creek reservoir and thence on to Broken Hill, a total distance of 71 miles. Of course, O'Connor could not know that Broken Hill would eventually adopt a pipeline scheme, but there would have been sufficient evidence in the 1890s of the water shortages in that district to have added considerable weight to his case, had he thought it necessary to refer to such evidence.

Finally, in O'Connor's defence, it might be asked how critical was his citing of rainfall figures—right or wrong—to the adoption of the pipeline scheme by the government. It was probably far from being the deciding factor. The weight of evidence of dire water shortages was surely convincing enough. Pressure on the government to adopt the scheme was amply supported by the heartbreaking stories in the press and by word of mouth of the general hardship, backed by the alarming and growing number of deaths from typhoid, the inadequacy of the existing dams, the failure of deep boring, and the overstrained capacity of the condensing plants. Forrest's impassioned speech in support of the Bill's second reading had no need of rainfall figures. He spoke from his own knowledge of the country:

> There is no doubt—I speak from experience—that there is in the Eastern portion of this colony a very small and uncertain rainfall. There is not a single drop of fresh water at Southern Cross unless it is caught from the heavens, or unless it is condensed, and if you catch all the rain it is possible to secure there will not be nearly enough.[18]

He then goes on to say how painful it was to see tired and jaded horses left on the roads to die in the wilderness. More than anything else, it was Forrest's stature as an explorer, his intimate knowledge of the conditions on the goldfields, and his genuine feeling for the suffering of the prospectors and their families that disarmed his parliamentary colleagues.

In the matter of the rainfall figures, the case against O'Connor remains, even if seen in the worst light, 'not proven', but the more likely interpretation is that he was innocent of any deliberate deception; that he was merely making a comparison—the word is used in the title of a section of his report—between the abundant rainfall in the Darling Range and the uneven, sparse rainfall at Coolgardie.

In spite of Blainey's argument, O'Connor's reputation and integrity remain secure; the accusations against him, provocative, interesting maybe, but unsustainable. They take no account of the upgrading and extension of the pipeline scheme to newly serviced areas in later years. O'Connor's estimate of the amount of water required on the goldfields at the turn of the century (5 million gallons daily) was indeed a gamble. It was based partly on the less than adequate information supplied by the mining companies themselves, but also on O'Connor's perception of the future demand: his pipeline was built, like his other works, not only for his time but also for the years to come. That he was intensely proud of his Goldfields Pipeline Scheme is amply proved by the large number of copies of his report that he sent around the world, not only to engineering colleagues but to his friends and relatives in New Zealand and Ireland. It is unlikely, having previous evidence of his honesty and punctiliousness, that he would have included in that report any deliberately false information. It is unlikely, too, that had there been any falsification, O'Connor's critics, both in and out of parliament, would not have recognized his error at the time and used it as ammunition against him.

The extant letters in the files acknowledging receipt of his report and praising its boldness are testimony to O'Connor's pride. He must have been aware that he was at the helm of one of the most exciting and imaginative engineering schemes then in existence. Surely an amount of professional pride in such circumstances is pardonable.

20

O'CONNOR CELEBRATED HIS 54th birthday with his family on 11 January 1897, a little over a week before his departure for Europe.

Although the daily paper did not publish a New Year's Honours List on 1 January 1897 as it does today, it is more than likely that the Chief would have already known of his award of Companion of the Order of St Michael and St George (CMG) and the intended investiture in London later that year. The 'most distinguished order' of St Michael and St George was created in 1818 by the Prince Regent (later George IV) and was bestowed mainly on those holding 'high and confidential office' in the colonies in recognition of outstanding service. The order was limited to 300 knights (among whom Sir John Forrest was one) and 600 companions. Clearly, it was Forrest himself who recommended to the Governor that O'Connor should receive the award in recognition of his work on Fremantle Harbour, then nearing completion.

O'Connor admitted later that during those months prior to his departure, his health was 'far from perfect' and much affected by overwork and anxiety.[1] The only photograph we have of him taken in 1897 is the official investiture portrait, which shows a proud, mature and upright man, in formal pose, perhaps looking a little older than his age (see next page). His head is now almost bald, a condition accentuated by his high forehead. He wears a bushy grey moustache and trimmed beard in the height of late Victorian fashion. His dark eyebrows—long a characteristic feature—shade his deep-set, penetrating eyes. Both in this photograph and in others, the eyes give an impression not only of the deep thought behind them but of a blending of curiosity with just a hint of amusement.

The Chief left Perth on the Albany express on Friday, 22 January, to board the Orient liner *Oraya* in Princess Royal Harbour, due to sail the following day

Official investiture portrait of C. Y. O'Connor, 1897.

Courtesy Muriel Dawkins

for London via Colombo, Suez and Brindisi. As he attempted to rest in his swaying first-class sleeping compartment on the overnight train journey, there must have been some satisfaction for him in the knowledge that his new harbour at Fremantle was nearing the stage when such tiresome train journeys would become unnecessary. One of his missions in London would be to accompany Sir John Forrest to a meeting with Mr Anderson, the director of the Orient shipping line, to acquaint him with the new facilities pending at Fremantle. As it turned out, the job of convincing the mail companies of the advantages of calling at Fremantle rather than Albany was to prove a long and difficult negotiation.

It is tempting to wonder how O'Connor, the compulsive worker, occupied his time as the *Oraya* steamed around Cape Leeuwin, the south-west corner of Australia, and headed north-west across the Indian Ocean. There was no ship to shore communication (Marconi's wireless telegraphy experiments were being conducted in Britain at the time but did not bear fruit until the following year), and there were none of O'Connor's beloved horses to ride. This would be a sea change, described by him on his return as 'a boon'.[2] But perhaps he allowed himself to gaze eastwards as the ship crossed the 32-degree parallel, the nearest point to Fremantle, which then lay over the horizon, and see, in his mind's eye, the thousands of tons of cargo and passenger shipping that would soon crowd those waters and make his harbour the western gateway to Australia. He had left the works there in the capable hands of the executive engineer, George H. Rice, and the engineer-in-charge, John A. McDonald. For O'Connor, by this time work on the harbour was virtually finished. Both the North and South moles had been completed; so, too, the southern embankment, where work was progressing on the reclamation of 54 acres. The northern embankment had also advanced. The rock bar across the river mouth had all but been removed, the rock being deposited on the north side of the North Mole. Dredging sand to a depth of 32 feet was in progress in the harbour basin, and the sand was being used in the reclamation. The south quay was being constructed, and the timber decking laid. One thousand feet of wharf was due for completion by the middle of the year, and the first ships to use the inner harbour were expected soon afterwards, while O'Connor was still away overseas.

It was an enforced holiday: there was nothing that O'Connor could do now but relax on deck and enjoy the voyage.

We know that O'Connor had last visited London in December 1864 on his way from Ireland to join his sailing ship for New Zealand. There had been great changes in the capital of the British Empire in the intervening thirty-three years and 1897, Queen Victoria's Golden Jubilee year, was a good time to

Fremantle Harbour, 1896. General view from Arthur Head;
dredging and reclamation in progress.

Courtesy Battye Library 28435P

Fremantle Harbour, 1897. Constructing the new south wharf (later named Victoria
Quay). Dredgers at work, right; rock blasting operations in the background.

Courtesy Battye Library 5079P

observe them. The extraordinarily fertile era of social reform, literature and industrial invention, eponymous with the Queen's name, was drawing to a close. The Promethean personalities of Victorian England—Tennyson, Newman, Manning, Carlyle, Darwin, Dickens and Disraeli—had departed the stage by the time of the Jubilee. William Gladstone, England's longest-serving Prime Minister, synonymous with the age and champion of Irish Home Rule, had at last retired. That year, the Grand Old Man was diagnosed with facial cancer, and died within twelve months, but not before the colonial Prime Ministers, including Forrest, had paid their respects by visiting him in his country seat, Hawarden, in North Wales. Shortly after O'Connor arrived in England, his compatriot Oscar Wilde (with whom, one suspects, he would have had little sympathy) was released from Reading Gaol to go into voluntary exile on the Continent. Gilbert and Sullivan's most popular works were written; their final opera together, *The Grand Duke*, was not a success and suffered a short run at the Savoy. Another compatriot, Bernard Shaw, having already staged and published his own first plays, was writing theatre criticism and praising Ibsen in the *Saturday Review*.

The sights and sounds of London, too, would have undergone major changes in O'Connor's absence. Gas lamps were giving way to electricity. The streets were more dangerous for pedestrians then than they are today, being choked with double-decker horse-drawn buses, four-wheelers and hansom cabs, among which the occasional motorcar could be seen in 1897: the early 4 mile per hour speed limit had been lifted the previous year. The bicycle was becoming fashionable for both men and women. Below ground, too, O'Connor would have experienced changes: he probably travelled on the first underground railways linking mainline stations and the city. Since the 1860s, clothing had become more drab and workaday: for women, the bustle had disappeared, replaced by long dark skirts and high-necked white blouses; and for men, the bowler hat was beginning to replace the topper, with the dark suit replacing the long frockcoat. Although Arthur Conan Doyle had supposedly killed off Holmes at the Reichenbach Falls four years earlier, and the detective was not to return until after O'Connor was back in Australia, in many respects this was still the London evoked in his novels: a London of gaslight, hansom cabs, and fogs that hid footpads, opium dens in Limehouse, and disreputable boatmen sculling on the river.

O'Connor would have had plenty of time to acquaint himself with London life and business since, as he later explained, he spent most of his eight months' absence in the capital in connection with the Goldfields Pipeline Scheme:

I found that the engineers were of opinion that the project was of
such vast magnitude and importance as to be deserving of the most
minute care…They went into the scheme with an amount of care
and precision, and to such an exhaustive extent, that it occupied a
great deal of the time.[3]

The Commissioners, as they were termed, were asked specifically to inquire into
certain aspects of the scheme, including the character of the pipes themselves,
their thickness and jointing; the reservoirs and their sizes; whether the pipes
should be duplicated, side by side; whether they should be laid in trenches or
left exposed on the surface; and details about the pumping stations, their size
and location. O'Connor was provided with an office next door to Sir John
Carruthers's in Westminster, and the Commissioners deliberated at the Institution of Civil Engineers in Great George Street. They did not necessarily meet on
consecutive days—the individual members had other commitments—and they
travelled to various parts of the country, often accompanied by O'Connor, to
inspect pumping machinery, steel plants and existing water schemes.

O'Connor was thus in London for Queen Victoria's Golden Jubilee
celebrations on 22 June and met Sir John Forrest who, as a colonial Premier, was
among the official guests at various functions including the service of thanks-
giving in St Paul's Cathedral. What O'Connor thought of the lavish festivities
we do not know. They were as much a celebration of Empire as they were of the
Queen's long reign: an Empire that *The Times*, writing of the event, described as
'the mightiest and most beneficial ever known in the annals of mankind'. The
Queen, seated in an open landau, was preceded by a magnificent, colourful
procession of 50,000 troops of the Empire, the largest military force ever
assembled in London. Among them were cavalrymen from New South Wales,
hussars from Canada, head-hunters from North Borneo, princes from the
Indian Imperial Service, and police detachments from Hong Kong, British
Guiana and Cyprus. As an expression of national pride, confidence and genuine
rejoicing, the spectacle was unsurpassed and universally unquestioned. Even
O'Connor's more quarrelsome compatriot Bernard Shaw, who generally
espoused provocative radical opinions without fear or restraint, on this occasion
was moved to compare favourably, in his review of a production of Ibsen's
Ghosts, the Queen's wisdom and knowledge of the common people with those
of the play's heroine, Mrs Aveling.[4]

Without seeing any inconsistency in his position, O'Connor would have
combined his love of Ireland with pride in the achievements of the British
Empire: achievements to which he had significantly contributed. It is likely he

took pleasure in witnessing the festivities and the part played in them by his Premier and friend, Sir John Forrest. A month later, on 21 July, in recognition of his colonial contribution, he was bidden to attend at St James's Palace for an investiture of companions by His Royal Highness, the Prince of Wales, on behalf of Her Majesty. On enquiry at St James's Palace, O'Connor was informed that 'Levée Dress will be worn'.[5]

Although the personal diary that C. Y. O'Connor, CMG, kept that year, which was until recently in private hands, is now lost, we know that on 23 July he bade goodbye to Sir John and Lady Forrest at the start of their return voyage to Western Australia. He had another month of crowded appointments. During this period, O'Connor managed to fit in a visit to his old country and stay with his surviving relatives, including his sister Frances. He had been away from Ireland more than thirty years.

At the beginning of August, he resumed government business, culminating in a visit to the Krupps works in Essen and other engineering establishments in Germany. He continued by way of Munich, Bologna and Brindisi, where he joined the ship *Oceana* on 22 August for his return to Australia.

Although the *Oceana* would berth at Albany on O'Connor's return voyage, a significant event had occurred in Fremantle while he was away that would spell the gradual decline of Albany as the major port of call in the colony. On 4 May, the large ocean-going steamer *Sultan* steamed into the river and tied up at the new berth on the south wharf. In its own small way, the occasion was marked with as much pomp and pride as the Jubilee celebrations were the following month in London. The Forrests had delayed their departure for London to officiate at the ceremony. As the ship left its moorings in Gage Roads and moved slowly towards the entrance of the river, watched by a cheering, waving crowd lining the entrance moles, Lady Forrest herself took the wheel (no doubt under the close supervision of the pilot, Captain McDonald).

> The brilliant spectacle created by the unusual display of flags and the majestic appearance of the large steamer at the spot so recently a sand-bank attracted crowds of people to the wharf during the day, and among the townspeople many mutual congratulations were exchanged.[6]

A lavish luncheon was provided for guests in the ship's saloon. One of the owners of the *Sultan*, Charles Bethell, welcomed the Premier and leading politicians, shipping executives, Fremantle councillors, harbour engineers and various dignitaries. He proposed a toast to Sir John and Lady Forrest and said it was a great honour to be connected with the vessel that had the grand distinction of being the first to use the new harbour. Sir John Forrest responded by declaring it a proud day for the people of Fremantle and the colony. Lest this be thought Hamlet without the Prince, C. Y. O'Connor was not forgotten. Forrest very generously paid him tribute. 'Everyone knew', he said, that 'the chief credit for the scheme was due to the Engineer-in-Chief'—a remark greeted with applause. Forgotten were the protests over noise and vibration due to rock blasting. Forrest said that he could not speak highly enough of O'Connor's conduct with regard to the harbour work that he had taken on, risking his reputation as an engineer by giving advice 'contrary to engineers of worldwide reputation'.[7]

The newspaper report ends by quoting the captain of the *Sultan*, F. Pitts, who stated that the entrance to the harbour was a good deal easier to negotiate than the one in Albany, and any ship-master could enter without a pilot. However, months of long and spirited negotiations were necessary before the P&O and Orient lines, the colonial postal authorities, would agree with Captain Pitts, and thus the *Oceana*, carrying the Engineer-in-Chief back to Western Australia, berthed as usual at Albany on 16 September.

The *West Australian* of the following day carried its correspondent's report of a lengthy interview with the returning Chief. It reads today as overly pompous and self-important, but the style is much in character with respectful newspaper interviewing at the time. O'Connor gave a detailed outline of the investigation conducted by the Commissioners into his Goldfields Pipeline Scheme. Perhaps anticipating that the newspaper would be eagerly read not only by the supporters of his scheme but by his critics, he seems transparently proud of the successful outcome of the investigation, quoting at length from the commission report that thoroughly endorsed the project:

> While the scheme, if carried out, will be the largest of its kind in the world, there is nothing in the nature of it, or in any of its details, which is in the least degree impractical or unprecedented. And with reasonable care and skill there is no reason to suppose otherwise than that it will prove to be entirely satisfactory.[8]

O'Connor may have felt that the ringing endorsement, by internationally recognized authorities, of the scheme as devised by him in his July 1896 report would smooth the way for the work to begin and silence the doubters. After all, was not his recommendation that the scheme be submitted to the Commissioners in the first place partly intended to accomplish this? But the way forward was not to be smooth, nor free of personal criticism, and, as we have seen, the result has not been universally accepted even today.

Throughout modern history, most epoch-making engineering works involving large sums of public money have been matters of bitter controversy and have generally spilt blood before they were finished: one only has to think of the Sydney Opera House and the Snowy Mountains Hydro-electric Scheme in New South Wales; or the West Gate Bridge in Melbourne; or the Ord River Dam in Western Australia; or the Aswan Dam in Egypt; or the Channel Tunnel in Britain. Only on completion of these projects, when their utility and beauty become self-evident, are the early controversies largely forgiven and forgotten.

When O'Connor arrived back in Australia in September 1897, he had some grounds for believing that his main worries might be over. But in reality, the year 1897 was only the conclusion of the first act in a drama that would end, for him at any rate, in tragedy.

21

SOMEWHERE BURIED AMONG O'Connor's copious luggage when he arrived back in Fremantle would have been his companion's medallion, a red St George's Cross set on a star of seven clusters, each with seven silver rays. In the centre was a depiction of St Michael encountering Satan surrounded by a fillet with the motto *Auspicium melioris aevi*. One can imagine the family hesitantly attempting a translation of the Latin, and being struck by the promise of the motto: 'A pledge of better times'.

The irony of that promise would not have been apparent late in 1897; on the surface, things had never seemed more auspicious for O'Connor. He had returned to Australia fit and refreshed both in body and spirit. His Goldfields Pipeline Scheme had been endorsed by the ablest professional authority, and his remarkable harbour was nearing completion and much praised. His standing among engineering colleagues both in Australia and in Europe had never been higher. The pipeline scheme was featured in engineering journals and trade papers around the world. The files testify to the large number of manufacturing companies writing—begging—to be allowed to be associated with the enterprise by providing equipment, and the many individuals offering advice and their services. Hodgson seems to have had the job of drafting most replies to these applicants, and one amusing note to his Chief informs him that

> We have a large number of applications from people who want
> positions as supervisors. A man who cannot do anything else thinks
> he is eminently fitted to look at someone else doing something.[1]

Letters from manufacturers and their agents were referred to the Agent-General in London or to the specifications contained in the Commissioners' final report.

A pledge of better times therefore might have seemed a good omen. But opposition to the scheme was rumbling away in the background and given encouragement by the difficulty the government was having in raising sufficient loan funds. A month before O'Connor arrived back in Australia, the respected Speaker of the Legislative Assembly, Sir James Lee-Steere, whose son was later to marry O'Connor's youngest daughter, attacked the scheme in a well-reported speech at Bridgetown. It did not meet with his approval, he said, and 'would cost considerably more than the estimate, and the time taken for construction would be greatly in excess of the time estimated'.[2] In October, amid squabbling in the Legislative Assembly, an attempt was made to have the whole question of supplying water to the goldfields submitted to a select committee.

Forrest stood his ground and so forcefully defended the scheme that much of the opposition was gradually subdued, if not entirely silenced. Even F. C. Vosper, the new Member for North East Coolgardie, declared dramatically that 'as a representative of the goldfields my voice will not be heard again in the House in opposition to it'[3]—a promise that famously he did not keep in the *Sunday Times*.

While O'Connor was away in England, his wish to be relieved of the management of the colony's railways had been granted—although not in a manner that would have given him much satisfaction. John Davies, the former traffic manager serving under O'Connor, had been appointed general manager in his place. O'Connor, as Engineer-in-Chief, would continue the practical work of design and construction of railways but he would lose financial control to Davies and Piesse. Thus, those in charge of business and traffic arrangements were deemed best able to determine the colony's railway building program. Both Davies and Piesse believed that O'Connor had 'erred on the side of economy' and that because the railways were now showing a profit, and there were funds in hand, a more liberal policy of spending should be introduced: new, more elaborate stations and buildings erected, and branch lines better equipped. O'Connor was, in a sense, now frozen out of these decisions and this undoubtedly caused friction between the Chief and Davies.

> It was well known that Davies and O'Connor did not 'take tea together' and relations between Piesse and O'Connor were also strained because the latter resented usurpation of the functions of the Works Department.[4]

O'Connor continued to offer opinions and advice, which Davies, in turn, resented. Eventually, their disagreements led to a complete breakdown of communications between them. O'Connor, Chief of Public Works, found himself in the demeaning position of having to act on Davies's decisions with regard to railway planning and construction. But O'Connor maintained that it was 'a fallacy that the officers who worked the railways were the best judges as to how they should be constructed'.[5]

O'Connor was to be proved right in many of his criticisms of Davies when it was seen that the strong position of the Government Railways that had been built up during O'Connor's rule was being dissipated under Davies. Four years later, in August 1901, Davies was charged with 'want of fidelity, capacity and diligence in his duties' and faced an independent board of inquiry into his management.[6] He was suspended from duty but later reinstated when all but two of the charges against him could not be proven. He was so unpopular by this stage, and so vilified in the press, that he found his position untenable and resigned to take up a position with the Midland Railway Company in England. Whatever O'Connor's differences had been with Davies, he could have received little comfort from Davies's public inquisition and the vicious campaign levelled at him in the press and in parliament—accusations that he was 'a tyrant and a cad' and that he was 'the most unpopular man in the State'. By mid-1901, O'Connor's own position was attracting criticism and he might have observed how the public appetite for blood was directed at once praised heads of departments and public figures—'tall poppies'—and how they were pilloried, often unfairly.

But on O'Connor's return to Western Australia in 1897, these railway squabbles lay in the future.

On the last day of the year, the O'Connors attended a garden party in the grounds of Government House. Doubtless the Chief would have worn his companion's insignia publicly for the first time. His daughter and son-in-law, Mr and Mrs Simpson, were also present, along with Eva, who would marry one of O'Connor's engineers within twelve months.

In the New Year of 1898, O'Connor and his assistants continued with their work of preparation for the pipeline in anticipation of the money being forthcoming. They must have been greatly heartened to be informed by the minister, Piesse, of the arrival of the Premier's telegram of 17 January, sent from Albany. Sir John was on his way to a Constitution Convention in Melbourne but was able to confirm that a second attempt to raise a loan had been successful.

> You may now proceed to put the Coolgardie Scheme in hand at once,
> to proceed with the railway to the dam site, to build the dam, to call

for tenders for pumps and pipes, and generally push on with the great work.

He ended his message by expressing the hope that they would all be with Mr O'Connor to see 'a river of fresh water on the Coolgardie Goldfields an established fact and a complete success'.[7]

As O'Connor had ruefully predicted, once the authority had been given for the start of the work, it was 'all hurry scurry'. In an interview published in the *West Australian* of 22 January, he confirmed that cables had been sent to the Agent-General in London, instructing Sir Malcolm Fraser to announce the launching of the scheme and invite tenders for the supply of 330 miles of pipes, each 28 feet in length and with diameters ranging from 26 to 31 inches. Similar advertisements were inserted in Australian papers. The Chief also stated that the work of constructing a tram line from Mundaring to the dam site on the Helena River would commence without delay and would be carried out departmentally. The decision to manage the work departmentally rather than by contract would give rise to controversy and attacks upon O'Connor and his department over the coming months. The *Morning Herald*, one of whose directors was O'Connor's old adversary Alexander Forrest, and which had supported the scheme back in October, now led the way in its support of handing the work over to private enterprise. In both its leaders and letters columns, the paper opined that although the Engineer-in-Chief might have been an expert in harbour building and planning railways, he 'has never had any experience in such undertakings as that under discussion'.[8] The *Herald* seemed much taken with a certain Mr Bargigli, a French engineer representing an English firm, who had travelled throughout the goldfields and had offered to construct the pipeline for £2,231,000. It was interpreted as a serious challenge. Bargigli's proposals were clearly set out in the *Morning Herald* and the *West Australian*, and attracted considerable publicity at the time.[9] He was received by the Premier on 5 April and defended his company's intention in both newspapers. But O'Connor was not impressed and advised his minister that Bargigli and others had not the advice and assistance that the Commission of Engineers experts in London had had:

> It was presumptuous that they should take it upon themselves to condemn in the most sweeping manner, the conclusions which the Commission arrived at after all this labour and study and I think it cannot be doubted that the advice of the Commission must be more valuable than that which has been volunteered by irresponsible outsiders.[10]

O'Connor's private opinion can best be summed up by a scribbled note on the bottom of one of his memoranda on the subject to Hodgson: 'Mostly bosh and special pleading'.[11]

In spite of the pressure brought to bear by powerful contractors and their friends in parliament, the work continued to be managed, and labour hired, by the Public Works Department. On 18 February, the Trades and Labour Council wrote to the minister thanking the Public Works Department for using day labour rather than contract, citing similar circumstances when building Fremantle's South Mole, which, it argued, proved more efficient and more economical.[12]

Active operations—confined to clearing the forest and laying railway lines—commenced at the Helena Valley dam site near Mundaring early in February 1898. The choice of site had been determined by Hodgson and his team after extensive exploration of some 3,000 square miles of the Darling Range. From thirteen possible sites, Mundaring on the Helena River was finally selected because of its superior catchment area of 569 square miles, capable of supplying, in one year's storage, two years' demand for water after taking into account likely evaporation. Hodgson set up an office in the area and a telephone line was provided between the site and O'Connor's office in Perth.

We get an idea of the relationship between the Chief and his second-in-command from the voluminous correspondence passing between them. The two engineers trusted and respected one another, O'Connor relying on Hodgson's specialist advice. But there is a sense in which Hodgson is cast as the junior partner. O'Connor was very much in command of the ship, being informed of all decisions and receiving and sending detailed information on every aspect of the work in progress. When Hodgson begins the job of drafting technical instructions for the surveyors at the start of their work in plotting the path of the pipeline, he complains to his Chief that Muir, the head surveyor, had been given certain instructions by O'Connor that conflicted with Hodgson's work. O'Connor writes back, somewhat apologetically, 'the instructions which I gave Mr Muir were of a merely general character in order to enable the preliminary work to be started'. And when the draft copy of the instructions are sent to O'Connor, he replies to Hodgson, 'I think these instructions are very excellent indeed'. Later he writes an additional note of advice to Hodgson: 'I made it a practise when I attended personally to the constructions of railways to walk or ride over the whole line and confirm or alter each and every proposed structure'.[13] This was the nature of O'Connor's management style: issuing a veiled instruction for Hodgson to do as he thought it should be done. O'Connor now commanded the largest staff of any government department: 300 at Mundaring alone, and also a large force still at work on Fremantle

Mundaring Weir, c. 1898. Early work in progress on the dam wall.
The temporary weir can be seen in the distance.

Courtesy Battye Library 209337P

Harbour, as well as draughtsmen, clerks, assistants and cadets in Perth. He was tireless in making numerous inspections of both sites, holiday periods not excluded. (On Boxing Day 1899—a holiday—he would even take his family to Mundaring on the train to inspect, with them, the work in progress.)

In April 1898, excavations for the weir were started and in August the railway to the construction site was completed, making transport of materials and personnel from Fremantle and Perth both rapid and comfortable.

In October, two key events happened in connection with the pipeline: the Coolgardie Goldfields Water Supply Construction Bill was passed, two years after the scheme had been adopted by parliament, and contracts for the supply of pipes were signed with the Australian engineering firms Mephan Ferguson and G. & C. Hoskins. With regard to the Bill, it was O'Connor himself who urged its drafting when he observed that no parliamentary authority had yet been given for the construction and the necessary resumption and purchase of land. Having convinced his minister of the necessity, he himself was invited to draft the Bill but objected strongly, declaring that he had no experience in drafting legislation and could not afford the time away from his other duties. The Bill was passed after several members had 'railed against the maddest scheme ever introduced into the legislature, and was sure to cripple the colony for a decade'.[14]

In the matter of the contracts for the pipes, the Australian firms' tenders were considerably lower than those received from manufacturers in Europe and America, at a price, delivered, within a few shillings per pipe of each other. Mephan Ferguson head office was in Melbourne and Hoskins in Sydney. O'Connor made an inspection of the Adelaide works in June 1898, and was soon convinced that the requirement was so exceptionally large that the output of at least two firms would be needed to complete the work on time.

The method of constructing steel pipes at that time, whether they should be welded or riveted and in what configuration, was a subject of much investigation and deliberation. Pipes could not be moulded in ready-made cylinders as they are today. They were generally formed from two separate lengths of steel sheeting curved in semi-circles, and either riveted or welded along the seams to form the tube. The Commissioners had first recommended that the pipes be riveted longitudinally, but there were problems inherent in this method: the likelihood of rust building up along the rivets, and friction from the protruding rivet heads impeding the water flow inside the pipe—nominally very little in one pipe but building up significantly over many miles. The story is told that Mephan Ferguson, the inventive and imaginative head of the Melbourne company, had long puzzled about this and similar problems with constructing pipes. Late one night in 1896, he was sitting at his desk with his son beside him. He opened a drawer and observed closely the dovetail joints, which suddenly inspired him with a new idea. He is reported to have struck the desk with a mighty blow and cried, like Archimedes, 'Eureka! I have got it, I have got it'.[15] He then determined to adapt the principle of the wooden dovetail joint for use in metal, and set about making the first prototype locking-bar joint: a strip of metal the length of a pipe formed into the shape of a horizontal H. The longitudinal flanged or beaded edges of the steel plates were then inserted into either side of the H and the arms closed over the flange so as to make a tight fit. O'Connor was so impressed with the 10-mile test of locking-bar pipes in South Australia, which could stand a water pressure of 400 pounds per square inch without any seepage of water, that he recommended their adoption for his pipeline. Moreover, this resulted in an overall reduction in the cost of the pipes and pumping equipment, as less energy was required to pump the water.

G. & C. Hoskins agreed to use the locking-bar method, and both firms set up works on the main railway line east of Perth. Mephan Ferguson established itself at Falkirk, renamed Maylands after Mephan's eldest daughter, May, while Hoskins operated from a site near Midland Junction. Steel plate for the pipes made by Hoskins was supplied by the Carnegie Steel Co. in Pittsburg in the United States, and the steel for Mephan Ferguson came from four different German steel companies.

As busy as O'Connor now found himself, he still made time to respond to various demands for his advice on less urgent projects. O'Connor's word was the ultimate engineering authority, a kind of imprimatur that, when delivered, sanctioned a course of action by a statutory body. As one example, on 8 January he accompanied the president of the Acclimatization Committee, Winthrop Hackett, on an inspection of the site of the soon to be established Zoological Gardens in South Perth, where he approved a proposal to sink an artesian bore. He also inspected both the Mends Street and Barrack Street jetties and recommended restructuring in anticipation of the crowds who were expected to use ferry transport for visiting the gardens.[16]

While initial work and voluminous administration continued apace on the pipeline scheme, significant events were also taking place in Fremantle. In February 1898, another ceremony was enacted, on this occasion to welcome the first mail ship into the port, the 6,600-ton German steamer *Prince Regent Luitpold*. The captain, H. Walter, was entertained by government ministers at a formal supper in the Fremantle Town Hall, at which the mayor and councillors of Fremantle were present and also the Engineer-in-Chief. Speeches were made in praise of the port, the shipping line, North German Lloyd, and government policy that had led to the building of the harbour. Councillor R. S. Newbold proposed a toast, 'Success to Fremantle Harbour Works', and O'Connor responded by reminding the assembled guests of the important role that the late W. E. Marmion, Member for Fremantle, had played in the building of the harbour. O'Connor admitted that they had 'often had little tiffs and differences as to how the work should be carried out', but said that 'Marmion had devoted himself without stint to the work, and when he [O'Connor] had arrived in the colony he found his way made easy'.[17] O'Connor, in deference to the nationality of the guest of honour, then paid tribute to the German nation and spoke of his cordial reception in Germany the previous year when he visited the Krupps steel works.

The centre of Fremantle was again the scene of an important function towards the end of the year, but a private one: the grand, fashionable wedding of the O'Connors' third daughter, Eva, at St John's Church on 7 December. Like the wedding of Aileen three years before, the occasion was attended by leading families and friends in government, including the Governor, Sir Gerard Smith, and Lady Smith, accompanied by their daughter, and Sir John and Lady Forrest. The groom, George Julius, eldest son of the Bishop of Christchurch, had come to Western Australia as an engineer in the Railways Department. 'His extra-ordinary flair for imaginative engineering quickly brought him to the notice of his chief'[18]—and doubtless to the notice of the Chief's daughter. George Julius, although never himself a gambler, is remembered in the family as the inventor of the automatic totalizator, based on an earlier idea for a foolproof voting machine that was never taken up. George was born in Norwich, England, and moved first to Ballarat when his father was appointed archdeacon there, and thence to Christchurch in 1890, where he attended the university. After the wedding at St John's, the guests were received at 'Park Bungalow', which the family had recently reoccupied after leaving 'Plympton House'. Once again, O'Connor is reported to have given a cheque to the bride and groom; the Forrests, another silver salt cellar. Bishop Julius and his wife were unable to travel from New Zealand for the ceremony but are reported to have given the married couple a chest of silver cutlery. Mr and Mrs George Julius resided in High Street, Fremantle, and later moved to Sydney, where George eventually became chief of the fledgling Commonwealth Scientific and Industrial Research Organization and was knighted for his 'outstanding contribution to the cause of science'.[19]

The year 1899, when O'Connor turned 56, was a turning point in the history of the Goldfields Pipeline Scheme: work at the dam site was so far advanced, with contracts signed and money allocated, that there could be no turning back. It was also the year when malicious criticism, aimed at O'Connor personally, began to appear in F. C. Vosper's *Sunday Times*. Carping opposition to the scheme, both in parliament and in the pages of the *Morning Herald*, had existed since O'Connor's report of 1896, but this was fairly easily deflected by Forrest and his supporters. In 1899, the character of the criticism changed noticeably, becoming an attack on the probity and competence of O'Connor himself. However wounding this must have been to both the Chief and his family, no amount of criticism at this advanced stage was capable of altering O'Connor's plans, nor of substituting an alternative.

Much of the criticism at this time was provoked by O'Connor's preference—supported by Piesse—for engaging day labour managed by the Public Works Department rather than using private contractors. Powerful contractors with friends in parliament denigrated O'Connor's methods, and, as O'Connor argued, managed to create discontent within his own work force, whose best interests he maintained he always upheld.

If O'Connor was angry and aggrieved on this occasion, he did not show these feelings in public. After all, this was not something new; ever since arriving in the colony he had waged an ongoing battle in defence of his work force against the interests of powerful contractors. But his sense of injustice and frustration in the face of powerful vested interests may be judged from a revealing letter, written some years previously, to his old New Zealand friend, John Lomas. Lomas was a much respected Methodist lay preacher, coalminer, and powerful union leader who had worked successfully to improve the status and conditions of working men in New Zealand society. The year before O'Connor departed the country, Lomas was a pivotal figure in the two month long strike that tied up New Zealand ports and involved some 8,000 union members—a strike during which, one suspects, O'Connor might have privately, if not publicly, sympathized with the strikers. In the letter, O'Connor begs Lomas to intercede for him by writing to labour leaders in Western Australia to defend his record in labour relations. He asks him to tell them the truth about the contractors' methods. He complains that he is being 'traduced and misrepresented by the so-called labour leaders of this colony' who were being led by

> an association of builders and contractors whose only object is their own individual aggrandisement, but who have, nevertheless, persuaded the labourers here that their interests, and the labourers interests, are identical!!

O'Connor then tells Lomas about the pressure being brought to bear on the government by the contractors in their fight against departmental labour, and in favour of work by contract:

> Such is the domination of these men, wealthy and influential, that the newspaper report of the deputation carefully omitted the explanations that I gave to the effect, to the best of my knowledge and belief, I was acting in the interests of the labouring classes and in accordance with the views of all the ablest writers on the subject, rather than against them.[20]

O'Connor then acknowledges the views of Rousseau, Marx, Sydney Webb and Bernard Shaw. He refers to his 'bitter experience' in New Zealand, where his convictions and sentiments had been at variance with the Conservatives'. The Liberal and Labour parties, he says, 'knowing little or nothing of what I had done and felt for them, were very lukewarm in their support, and allowed me to be frozen out of the [public] service'. He tells Lomas that he had always acted in 'the best interests of the labouring classes, and had already procured for them the eight-hour day which they had vainly struggled for'.[21]

The letter is a remarkable cri de coeur from a man who has left us few documents revealing his inner feelings. Here is evidence of his frustration and his political beliefs. He believes he is a man grossly misunderstood, a man wronged. The style of the letter leads us to wonder whether he might be exaggerating, that he might be a little too pained, too self-pitying, and yet the tone is entirely consistent with the principles he followed throughout his working life: principles that had their roots back in Ireland when he was a child on his father's farm among the starving peasants; principles instilled by his evangelical upbringing, and later, by his struggle to make his way working under contract and uncertain employment in remote corners of rural Ireland. The letter to Lomas is also important because it explains more clearly than anything else the foundation of the growing antagonism towards O'Connor waged by those with more insular, sectional interests.

If we take as an example of this antagonism just one of the debates in the Legislative Assembly attacking the Goldfields Pipeline Scheme, that of September 1898, it is plain that O'Connor's views have the ring of truth. George Leake, J. J. Holmes, Frank Wilson and others proposed that the work be managed as a commercial operation. The tall, moustached Holmes, who led the attack, had a contemporary reputation as being 'hotheaded and never shirking a big deal if there was a chance of securing profit'.[22] He maintained that there had been 'bribery all round' and that the Engineer-in-Chief was deeply implicated. Frank Wilson, a shrewd Scot who had 'a nature as hard as Aberdonian granite',[23] called the scheme the wildest ever undertaken by the government and said it should have been abandoned twelve months ago. He was a director of Collie Coal Mines and perhaps harboured a grudge against O'Connor for his earlier criticism of the quality of the coal mined there. Leake, known at other times for his easy wit, said, 'Nothing would give me greater delight than to see the thing smashed up'[24]—ominous words when it is remembered that Leake was to become Premier in the last year of O'Connor's life.

No letter is to be found in the files to show that Lomas responded to O'Connor's plea for help, but even if one was sent, it would surely have had little effect on the mounting campaign by O'Connor's enemies.

The pipeline was set aside for a few days in January 1899, when O'Connor made one of his country trips to Bunbury, Collie and Busselton. He inspected the Bunbury harbour works on the 17th and left by special train the following day for Collie, where he inspected the coalpits and quarries. He commented on facilities at Brunswick Station in his notebook and suggested improvements. On the 19th, he took a special train to Donnybrook and Bridgetown for the purpose of inspecting the new line, returning to Bunbury overnight. The following day, Saturday, he inspected the jetty at Busselton and remained there over Sunday before returning to Fremantle, arriving there mid-afternoon on Monday, 23 January.

The steel plates for manufacturing the pipes began arriving at the new Fremantle port on 4 April, confounding the warnings of those critics who had said that they would never be delivered on time. Perhaps the pessimists were compensated by two entirely unforeseen mishaps that seriously slowed progress. The first of these was the discovery, during excavation at the dam site, of a fault line or fissure in the rock foundation. O'Connor was called to the site and took the responsibility of deciding that the dam could not be shifted; the fault and the loose rock had to be drilled and excavated to a depth of 90 feet below the river bed. He also recommended installing electric light so that two shifts could work, one by day and the other by night. Even so, the total operation to repair the fault delayed the start of the concrete construction of the dam by twelve months. This was not missed by the vigilant *Sunday Times*, which reported the delay with undisguised glee: 'The progress at present being made at Mundaring seems to indicate that the first half-pint of tarry fluid will reach the goldfields somewhere about AD 2000'.[25] The paper lampooned the engineers for choosing the Mundaring site and charged that a grievous blunder had been committed.

The second incident that year that frustrated progress on the pipeline, and involved a tragic loss of life, was the wreck of the *Carlisle Castle* in a gale southwest of Garden Island on 11 July. It was carrying a shipment of locking-bars manufactured in England and went down with all hands.

Progress on Fremantle Harbour was not similarly thwarted. The government had approved an extension to the North Mole by another 1,200 feet, which O'Connor had recommended. The *West Australian* was able to report, with a modest headline, on 22 May that the scheme was nearing completion. Another colourful ceremony of some significance took place there on Saturday afternoon, 20 May, with the launching of the 400-ton hopper barge *Platypus*, the first vessel of any size constructed in the harbour—or, more accurately, assembled from parts manufactured overseas. The event was marked with the customary assembly of Fremantle dignitaries, members of parliament and a brass band. Afternoon tea was served in a temporary structure up river adjacent to the launching site. Lady Forrest performed the traditional ceremony, breaking a bottle of champagne over the bow, but the significance of the event for us is more to be found in the speech made by the Engineer-in-Chief on that occasion. Replying to a toast by Mr R. B. Campbell, the locomotive engineer responsible for construction and acting chairman of the proceedings that day, O'Connor stated in the clearest terms what many historians have since come to believe, that his work on Fremantle harbour was 'the work of his life to which he had devoted all the thought he could bring to bear on the subject'.[26] O'Connor went on to thank his colleagues by name and pay tribute to the engineers who had worked on the harbour and had given him loyal and valuable support. And then he made the crowd laugh by referring to his somewhat singular appearance and his preferred mode of dress. He told the story of how Sir John Forrest, speaking at a previous function in Fremantle, described how he had selected the Engineer-in-Chief on recommendation without first having seen him. According to O'Connor, a lady listening to Sir John on that earlier occasion had commented afterwards that O'Connor surely could not have been appointed on his personal appearance alone. This was greeted with much knowing laughter because not only were O'Connor's physical features distinctive—his above-average height and the shape of his head—but he seldom dressed in a conventional manner, preferring large, wide felt hats to cover his misshapen head, and lighter, more dapper clothes than the sombre black thought appropriate among the professional classes.

The praise given to O'Connor and his engineers that day did not impress the editor of the *Sunday Times*. Later that year, he called for an investigation into the works at Fremantle, stating that they were 'alive with discontent and jobbery'. In broad terms, it was an attack once again on the use of day labour rather than contractors. The writer, quoting a discontented employee, stated that the slipway was 'botched', at a cost varying between £50,000 and £70,000, but could have been built with contract labour for £6,000. The attack, aimed at O'Connor, was couched in extravagant terms:

> All that O'Connor knows about engineering could, without crowding, be stated in a very small book and one consequence of that unfortunate circumstance is seen in the fact that his trusted subordinates are never likely to outshine the central luminary.[27]

O'Connor himself was unlikely to be shaken by these wild and inaccurate detractions, but their persistence over the coming months—like all media rumours and innuendo today, and like water washing away the stone—had the effect of undermining his high reputation and raising doubts, not only in the more susceptible public minds but in the minds of those politicians who were only too anxious to cut him down in size.

C. Y. O'Connor in characteristic dress and pose.

Courtesy Judge V. J. O'Connor

22

THE ARGUMENT ABOUT whether a new century commences with the last year of the decade or the beginning of the new one is familiar to us, we having witnessed the arrival of a new millennium, but such precision did not much trouble West Australians in January 1900. For them, judging by the views expressed in the newspapers of the colony, a new century and new hope dawned on 1 January. The people welcomed the newly born twentieth century with optimism and celebration, and what modern pedant would care to argue with them? The year 1900 really did usher in a new era marked by significant historical events that effectively changed Australians' perception of themselves forever.

For the first time in history, West Australians were engaged in a war overseas. More than 2,000 men, out of a total of 16,632 for the whole of Australia, had volunteered to serve with the British Army in the Boer War in South Africa. The progress of that war, dispatches from the front, obituaries of those killed and news of the wounded were fully reported each day in the newspapers. But claiming even more news space than the war was the coming referendum on federation, due to be held on 31 July. Political speeches on the subject, letters from readers, and column after column of editorial opinion were of such persistence and character as to invite an uncanny parallel with the debate on the republic referendum that took place close to 100 years later.

We do not know for certain what O'Connor felt about the question of federation, although it is surely safe to assume, given his visionary concepts in other matters, that he would have voted for a united commonwealth. However, we do know something about his attitude to the South African war, if only from a handwritten, incomplete draft of a letter to an unnamed friend, or perhaps to a public figure of influence, that survives in the archives. Although almost illegible, a mere scrawl, it begins, 'For the love of God and St. Patrick I pray you think of some means of avoiding involvement…', and asks whoever is being addressed 'to call on your councils, as you have called into your armies to seek a means of rejecting despotism'. The scribble ends by asserting that these must be the feelings of all enlightened Irishmen and West Australians.[1]

Both the South African war and the federation referendum filled so much space in the newspapers during the first half of the year that, apart from notable exceptions in the *Sunday Times*, little attention was given to the Goldfields Pipeline Scheme. It was an uneventful period, which even the Engineer-in-Chief so described in a letter to Sir Malcolm Fraser in London when the latter wrote asking for a progress report. The tedious work of repairing the fissure in the bedrock at the dam site was at last completed in January, and the monotonous but crucial survey of the pipeline route was well in progress. Photo opportunities of trains loaded with pipes and gangs of men laying and caulking them in freshly dug trenches were still in the future. Much thought was given to requests from Coolgardie to begin work on storage dams ahead of time in order to relieve mounting unemployment on the goldfields. O'Connor was sympathetic and asked Hodgson to prepare a schedule of those works that might be anticipated without endangering the success of the project.

A contract was let in March to James Simpson and Co. of London for the Worthington pumps, twenty complete sets to be delivered commencing the following year, at two-monthly intervals. The total cost was £241,750.

Contrary to the original advice given by the London Commissioners, O'Connor decided to lay most of the pipes in trenches, taking into consideration not only the intense summer heat and winter cold, but the extreme rise and fall between night and day temperatures. By the middle of the year, the first 30 miles of trenching had been excavated. Pouring concrete for the weir wall began early in the year, and continued day and night until completion in June 1902. A total of 77,508 casks of raw cement were imported for the purpose from Germany and Britain.[2] On this last matter, the *Sunday Times* conducted a dramatic scare campaign, alleging that the city of Perth was in danger of being engulfed by a wall of water due to the use of rotten cement in the dam wall. With what amounted in those days to a banner headline, it announced 'The Coming Destruction' and quoted an employee at the dam works 'who naturally did not wish his identity to be revealed'. The highly coloured prose foretold that a mighty flood would brush aside the dam 'like a piece of paper', sweeping all before it, and would eventually reach the city 'bearing with it all kinds of wreckage'. The reason for this disaster, according to the paper's informant, was a mixture of bad cement with good, lowering the overall quality. And the paper strongly hinted at corruption among persons 'not quite unknown to political fame'.[3]

Towards the end of the year, the same paper conducted a campaign over several weeks attacking the quality of the steel pipes with regard to both the protective coatings and the type of steel being used. Once again, it used information supplied by a correspondent, in this case an engineer named A. F. Smith. It was alleged in some detail that 'faults were plainly visible to the eye of any intelligent person, except the Engineer-in-Chief or the Engineer for the Coolgardie Water Scheme'.[4] Various statements by O'Connor in the *West Australian* rebutting these extravagant claims were probably missed by the majority of *Sunday Times* readers.

As improbable and amusing as they might seem to us now, such attacks had a cumulative effect on credulous minds at the time, arousing suspicions— perhaps hardly conscious at first—that O'Connor and his engineers on the Goldfields Pipeline Scheme might not be as honourable as had been supposed.

The editor and part-owner of the *Sunday Times* who was credited with publishing the stories was the fiery, crusading journalist and politician F. C. Vosper. He might be described today as a left-wing political activist, a republican who loudly supported votes for women, a minimum wage, penal reform and the rights of the working man. He was an implacable enemy of John Forrest and his government, including those like O'Connor who were seen to be part of the powerful Establishment. Vosper's tall, lanky figure and shoulder-length black hair, and his past, which included a three-month prison sentence in Victoria for inciting a riot, fuelled animosity and made him many enemies. As often as he may have been right, he was just as often wrong, and, like an undisciplined pugilist, he lashed out in all directions on the flimsiest of evidence. A courageous if erratic man, he died suddenly of neglected appendicitis on 6 January 1901, greatly mourned by a loyal following. At his death, even Forrest, so often the prime target of his venomous pen, acknowledged Vosper's courage and reforming achievements. Legend has it that Vosper's defamatory writings in the *Sunday Times* led to O'Connor's demise, but this can hardly be true. Vosper died fully fourteen months before O'Connor, and those accusations in the paper considered the most libellous were published well after Vosper's death. Among the many wreaths laid on the coffin at Vosper's funeral was one from C. Y. O'Connor: evidently he bore him no ill-will and may even have applauded some of Vosper's humanitarian campaigns.

Six months previously, on 31 July 1900, West Australians voted by a majority of two to one to enter the Australian federation. While this may not be judged a critical event in the personal life of the Engineer-in-Chief, federation was, nevertheless, a turning point leading to the departure at the end of the year of O'Connor's main ally and defender, Sir John Forrest.

Three days after the referendum, a less trumpeted but important letter was signed in London by the Imperial Postal Authority, finally agreeing to the use of Fremantle Harbour rather than Albany as the port of call for British mail ships in the colony. Since the opening of the port to large steam vessels two years previously, frustrating negotiations and arguments had been necessary to convince the postal authorities and the shipping companies of eastern Australia of the wisdom and safety of using the new port. Although the people of Fremantle had grown accustomed to the sight of foreign cargo vessels at the town wharves, the prize, the affirmation eagerly desired, was the far greater frequency and importance of P&O and other British mail vessels. Consequently, the arrival of the first P&O ship the *Ormuz* on 13 August 1900, followed by the *India* on 20 August, was witnessed with great rejoicing and festivity. Crowds gathered on the quays and the moles to watch the vessels' approach, setting a Fremantle custom that would survive for a hundred years and more. According to a newspaper report, the 'splendid liner' *India* was quickly overrun with visitors eager to explore the vessel, 'notices such as "Do not enter the cabin", or "Keep off the Bridge" only serving as incentives to do exactly the contrary'.[5] It is hardly exaggerating to say that these events in Fremantle marked the pinnacle of O'Connor's career. For this was the recognition, after nine years' toil, that Fremantle was now the international gateway to the colony and a major port in Australia. Sir John Forrest acknowledged this fact in his speech at the celebratory banquet on board the *India*. 'It was a proud day for the people of Fremantle', he said, 'and a proud day for the colony. It was a proud day', he added significantly, 'for my friend, Mr O'Connor'.[6] Unlike at previous Fremantle Harbour celebrations, the Engineer-in-Chief, although present, did not make a speech.

Also present at the banquet was O'Connor's minister, F. H. Piesse, his last public appearance before his resignation on 23 August. At a farewell reception for his Public Works Department staff, he paid tribute to their hard work and loyalty during his four years of office. He was known as a conscientious man who made no attempt to be ingratiating. In speaking of the Engineer-in-Chief, he said that he was thankful for the assistance he had rendered

> and although at times, as is inevitable, there have been one or two differences of opinion over some little matters, still it has never interfered with our friendship, nor the cordiality of our official relations.[7]

O'Connor, too, spoke of the cordial relationship that existed between them. Although Piesse had never been as capable as Venn, O'Connor may have been genuinely sorry to see Piesse go. Piesse was known to be 'generous to a fault'[8] and Forrest said of him in parliament that 'There is no man who ever occupied office who has devoted himself to the duties of that office with greater zeal and a greater desire to do right'.[9] His successor, who occupied the portfolio for less than a year, was B. C. Wood, a Perth auctioneer with the reputation of 'not having the ability or energy to do any harm'.[10]

On 16 November 1900, O'Connor engaged in his first public relations exercise connected with the Goldfields Pipeline Scheme. More than 100 parliamentarians and other guests were conveyed by a special train to Mundaring to inspect work on the dam. Led by the Engineer-in-Chief, the resident engineer, W. Leslie, and engineer-in-charge of the scheme, William Hodgson, the party was shown over the site and entertained to lunch in a pavilion especially erected for the occasion. The excavation for the foundations of the weir had been completed and the members of parliament would have been able to observe the lower sections of the concrete dam wall in place.

As the New Year of 1901 arrived, two major events dominated the news, to the exclusion of O'Connor's public works. The first, not unconnected with

C. Y. O'Connor, with parliamentarians and guests, at Mundaring railway station during the first public relations exercise connected with the Goldfields Pipeline Scheme.

Courtesy Battye Library 3214B/20 (213 269P)

the second, was the proclamation of the Commonwealth of Australia by Queen
Victoria on 1 January. A facsimile of the historic document was reproduced in
the newspapers, along with a message from the dying Queen wishing 'her loyal
and beloved subjects in Australia peace, prosperity and well-being'. For several
days, the festivities, official dinners and special events continued throughout the
colony and the country as a whole. 'In all its history Perth had never given itself
up so thoroughly and completely to the enjoyment of the hour', enthused the
West Australian.[11] The city was extravagantly decorated and thousands swarmed
there from all over the State to join in the festivities. According to one report,
in spite of the intense heat, fully 10,000 people gathered on the slopes of the
Esplanade to hear the proclamation read, and they all stood up and uncovered
their heads when the choir began to sing Handel's Hallelujah Chorus. Were
O'Connor and his family among the VIPs? We do not know for certain, but
surely they would have been among those invited.

The celebrated proclamation of the Commonwealth of Australia was likely
to have been one of the last royal documents to be signed by Queen Victoria.
The rejoicing of the New Year was soon afterwards followed by the news from
Osborne House on the Isle of Wight, on 22 January, that 'the longest regnant
of the sovereigns of England had breathed her last'.[12] The Victorian age of
unprecedented peace, social reform and industrial progress, with which
C. Y. O'Connor was so clearly identified, had come to an end. The Queen
Empress's reign had spanned the entire lives of most of those then living, and
her death was as much mourned throughout Australia as it was throughout
Britain and the rest of the Empire. She was, opined the *West Australian,* 'the best
loved and the most venerated of sovereigns'.

For O'Connor, who, as a leading government servant, must have attended
celebrations connected with the proclamation of the Commonwealth and later
would have signed the condolence book in Government House for the dead
Queen, it was business as usual at Mundaring Weir. On 7 February, he was again
conducting a parliamentary delegation to inspect the site of the dam. On this
occasion, the new Federal Minister for Trade and Customs, the Honourable
C. C. Kingston, together with Sir John Forrest, were important guests. The
occasion was to be Forrest's last public engagement before he resigned office to
enter the new Federal Parliament. His speech at the official luncheon, once again
held in a pavilion on the site, was a heart-felt eulogy to the work that, in a sense,
was as much his own as it was O'Connor's. Forrest reminded those present that
this would be the last occasion on which he would visit the great work at
Mundaring. He had been present at the birth of the project, he said, and had
protected it in its infancy from its enemies; now he was handing it over to

someone else. There was nothing he regretted more in handing over the premiership, he continued, than that he should not see the great work consummated. In an oratorical flight of biblical fancy, he said he felt as Moses did on Mt Nebo as he looked down on the promised land and had been refused permission to enter.[13] Suitably impressed, no doubt, the delegation returned to Perth by train, arriving at 6.00 p.m. Forrest's imminent departure for Melbourne to join the new Federal Parliament marked the close of his uninterrupted ten years of premiership, leaving, if not exactly a vacuum, certainly a period of instability and ministerial impermanence that was to have a deleterious effect on O'Connor's Goldfields Pipeline Scheme, and ultimately on O'Connor himself.

Saturday, 6 April 1901, was the occasion of the fourth Harbour Works employees' annual picnic at Bicton (probably that part of Bicton known today as Point Walter, a favourite picnic spot). The minister, B. C. Wood, and the Engineer-in-Chief and his wife were among those present. The photograph we have of the group (see opposite), which includes O'Connor, then an ageing figure in a white suit with hat pushed back from his forehead, is probably the last taken in which he appears. The traditional works outing included sporting events for the men and their families; for wives and children, novelty races, swimming and afternoon tea. In the tea tent, speeches were made and O'Connor took the opportunity of answering the critics of his labour policy. He said that the cost of dredging and the removal of the rock bar at Fremantle was as low as anywhere else in the world and 'had beaten all records'.[14] Did O'Connor know in advance that in the *West Australian* on the Tuesday following, the first shot in a prolonged war aimed at the day labour policy would be fired by his old adversaries, Messrs Atkins and Law, on behalf of the Contractors Association?

The letter, two whole columns in the newspaper, is important because it raised questions about the administration and fiscal management of the Public Works Department (and therefore of O'Connor himself) that would reverberate around parliamentary and newspaper corridors throughout the coming months. Couched in polite and seemingly rational language—which O'Connor later acknowledged—the writers' analysis of the comparative costs of contract versus departmental labour on recent railway works, harbour works and the pipeline was detailed and persuasive. Each of the works is considered in turn: that which focuses on the Fremantle wharf is a typical and sufficient example:

> Thousands of pounds have been spent in repairing and holding up
> the south river wharf since its erection which was very expensive in
> the first place. This is partly owing to the piles not being sufficiently
> driven, and because of bad workmanship. When this structure was

Harbour Works picnic at Bicton, 6 April 1901—possibly the last
known photograph in which O'Connor appears.

Courtesy Battye Library 5054P

being criticised in Parliament, Mr Gregory asked: 'who built the
wharf?' But not one of the Ministry was man enough to say that the
Department had built it under a manager and that no contractor had
anything to do with it. This wharf cost more than £16,000 to build
than the standard contract rates...even the bolts in the new work
show a lack of ordinary care, hundreds of them being too long. The
nuts are screwed to the end of the thread in green timber, leaving
nothing to tighten up after shrinkage.[15]

The writers, Atkins and Law, cite the success of contract work in the manufacture
of pipes for the Goldfields Pipeline Scheme, resulting in a saving of hundreds of
thousands of pounds. But they point out that instead of the saving being
deducted from the loan schedule, it had been absorbed elsewhere to make good
less successful areas of the project. With a litany of similar examples, backed with
figures, the contractors' case must have seemed conclusive.

Two days later, O'Connor's reply, of similar length and in similar detail,
was published. He opens by saying that although Atkins and Law had fairly and
reasonably put one side of the argument, nevertheless there was another side. He

answers each point in turn with similar conviction, and on the matter of the construction of the moles restates his central argument:

> I may say, as I have often said already without serious refutation, that whatever may be thought or said as regards the possibility of the work being done by a contractor at less cost than the Government have done it for, I feel absolutely certain that no contractor of stability would have tendered to do it for the price that the Government has done it for at the time when it was first undertaken.

O'Connor then, with a touch of irony, reminds his readers that in the matter of constructing moles, there might be benefits accruing from having the work done by disinterested officers of the government because

> a contractor who is making a large profit by tipping stone into the sea and has a personal and pecuniary interest in it [might] have a very vital and strenuous objection to tipping more stone into the sea than was actually necessary.[16]

The battle raged on with a second letter of similar length in the *West Australian* dated the following Monday, signed by Atkins and Law on behalf of the contractors. O'Connor's second response is even longer, taking the form of an interview with a reporter. He considers each of the public works in turn and shows that the arguments and the figures quoted by his critics are false. The reply is of special interest because O'Connor ends by summarizing his equivocal attitude to department versus contract labour:

> I am not at all wedded to departmental work exclusively, as against contract work, and would in fact prefer contract work where the conditions are such that the exact amount of work to be done, can be accurately defined, and where the plant required is of moderate cost, but I do not think that that applies to works such as Fremantle Harbour or the Coolgardie water supply, or to the construction of railways to the goldfields.[17]

Hardly had O'Connor finished dealing—more or less successfully—with the first wave of serious attacks on his works policies when a second controversy erupted in mid-1901 that was to cause him greater anxiety and embroil him in a prolonged media inquisition.

The locking-bar system of pipe construction had solved one major problem but there remained the question of joining one 28-foot length of pipe to another so as to render it permanent and water-tight. Traditionally, this had been done by hand-caulkers packing the joint with lead or pitch, a time-consuming and labour-intensive process. In addition, there was occasional damage to the pipes, and uniform standards could not be guaranteed. A Victorian contractor, James Couston, and his partner, James Finlayson, had earlier invented a machine that, it was claimed, simplified the caulking process and would reduce labour costs. Both Hodgson and O'Connor were impressed with the prototype and encouraged Couston to fully develop it, eventually recommending purchase of the patent by the government for the sum of £7,500. (The price paid would later become a critical factor in the 1902 Royal Commission, wherein it was shown that Couston would have happily accepted a lower figure.) A number of these machines were manufactured, and Couston himself was contracted to supervise and train men to work them. Therein lay the first of O'Connor's problems. The machines were complex; learning to use them required patience and hard-learned skills. The hand-caulkers resented the

Work on jointing and lead caulking for the Goldfields Pipeline.
Courtesy Battye Library 209357P

new methods and there were various teething problems—not least, delays in the supply of supporting equipment, and inexpert handling. Rumours persisted, fanned by discontented hand-caulkers, that the machines did not come up to expectations and were therefore judged a failure.

O'Connor answered the critics in the *West Australian* of 7 August. He agreed that there had been some difficulty in starting the process, but that then 'it was going quite satisfactorily and had been doing so for quite some time'. He confirmed that between 14 and 15 miles had been caulked and the only reason more work had not been done was because of delays in obtaining the ancillary equipment. He said that he had been assured over and over again by men who had seen the caulking in progress that the joints were better and able to withstand a greater pressure of water than hand-made joints. He quoted Hodgson, the engineer-in-charge, as saying that the machines were 'a splendid success'. On being questioned about the royalty payment to James Couston, O'Connor explained that only a portion of the fee, £2,700, had been paid as a deposit; the remaining £4,800 would only be paid when 12,000 joints had been made to the complete satisfaction of the department.

O'Connor's public statements did little to stem the flow of criticism and innuendo. Three days later, Mephan Ferguson added his assurances to those of O'Connor, stating in more technical detail the efficiency of the caulking machine, and O'Connor said:

> I have reason to be grateful, and the State too, to Mr Couston for his ingenuity in designing such an efficient caulking machine, and all the other appliances which he has devised in connection therewith.[18]

Assurances like these had little effect. Three days later, another attack on the use of the caulking machine was given prominence, and this in turn was answered by a joint letter from Couston and Finlayson on 19 August.

Behind O'Connor's public utterances in defence of the caulking machines was his own growing feeling of unease, a realization that progress on the pipeline was not proceeding at the rate required, and that the caulking machines—or at least the handling of them—may have been partially responsible. Couston admitted as much and argued that he was being hampered by public service red tape and restricted by working conditions, and that he could complete the caulking faster and more efficiently if he could employ his own men in a contract situation. This, of course, ran contrary to O'Connor's earlier stated policy. A change in direction, however, was foreshadowed when Walter Kingsmill, recently appointed the new Minister for Works by Premier George

Leake, indicated a more flexible approach to the question of contract work when he attended a demonstration of the machines at Cunderdin in company with the engineers in August.[19] Perhaps encouraged by Kingsmill's views, Couston wrote to the Engineer-in-Chief on 23 December formally proposing that his firm, Couston, Finlayson and Porritt, be contracted to finish the work of pipe-laying independent from the Public Works Department. Couston undertook to complete the pipeline not only faster but more cheaply than was possible under the then existing conditions:

> [We] believe that the work, if carried out by contract, would not only relieve Mr O'Connor personally of a great anxiety, but a very great service would be rendered to the Government and State generally... we are also prepared to recognise the eight-hour day principle and to pay at least the same rate of pay as is being paid by the Government.[20]

O'Connor pondered the matter over Christmas 1901 and in the New Year would seek Hodgson's advice.

The electric caulking machine, the cause of much controversy
when it was first brought into use.

Courtesy Battye Library 209356P

What followed in the early months of 1902 from O'Connor's change of course—entirely consistent with his earlier statement about not being wedded to contract work exclusively—would unleash in both Houses of Parliament, ably assisted by the newspapers, such an unexpected, undisciplined and vituperative attack on the Goldfields Pipeline Scheme, coupled with a denigration of the character and probity of the Engineer-in-Chief himself, that some tragic result would seem, in hindsight, to have been inevitable. It was as if that one administrative decision was just what O'Connor's critics had been waiting for. Now they had found the excuse to light the bonfire, the flames of which would ignite O'Connor's final breakdown.

23

THERE WAS NO sign of the coming storm in the first two weeks of 1902; rather, the opposite was the case if a glowing report in support of the Goldfields Pipeline Scheme in the *West Australian* of 12 January is to be believed. The reporter, in a confident opening sentence, stated that 'Today there are no opponents of the Coolgardie Water Scheme'. He went on to explain that while only a short time ago there had been many who belittled the work, those same people now believed that 'it was one of the boldest and most beneficial that has been evolved and carried out in the Southern Hemisphere'. Fulsome praise was heaped on those working on the scheme, and one by one the criticisms that had been levelled from time to time were shown to have been disproved by the 'substantial progress that was being made'. It is more than likely that Winthrop Hackett himself was the author of the piece, being a close ally of Forrest's. And although they were not quoted by name, it is possible to see in the report the hand of O'Connor (or Hodgson) as the source of the information given to the writer— the argument, for example, that the work on the scheme had properly only been in progress for two years, and that 'financial reasons had vigorously prosecuted against the work being operated until then'. Also described were the trenches, 'excavated almost as far as Coolgardie', and it was noted that even at that date there was 60 feet of water in the Mundaring Weir ready for pumping. In a clear reference to the *Sunday Times*, the alarmist predictions of disaster by 'these self-constituted engineering authorities' were ridiculed. The report ends by confidently predicting that 'within twelve or eighteen months five million gallons of water will be daily delivered at the goldfields and the intervening districts'.

Although this last statement turned out to be an accurate prediction, the writer was certainly deluded in his opening sentence; the silence of opponents may have been due to no greater cause than parliamentary Christmas–New Year holidays.

However, there was little holiday opportunity for the Engineer-in-Chief who, immediately after Christmas, asked Hodgson for his opinion of the Couston offer and received his reply on 11 January, O'Connor's 59th birthday. The engineer-in-charge of the scheme was in favour of the contract on condition that certain reductions in Couston's costings were made, and that a guarantee was obtained that the work of pipe-laying would be completed as far as Cunderdin by 31 March and as far as Coolgardie by 30 September. He ended by stating that, if contracted, 'the work could be carried out with greater expedition and satisfaction, and the Department would be relieved of much responsibility in regard to delays and mishaps'.[1]

Hodgson's endorsement was clearly instrumental in settling the matter in O'Connor's mind. The next step was to consult C. H. Rason, in company with Hodgson. It could not have been easy for the Chief and his second-in-command that Rason was O'Connor's third Minister for Works appointed within twelve months and had been only three weeks in office. He was known as a careful man, an able debater, his 'fluency and sarcasm dreaded by his opponents', but an educated gentleman 'even popular with his political foes'.[2] Understandably, therefore, Rason hesitated to accept the proposal and asked for a detailed memorandum justifying O'Connor's change of heart. This was drafted the same day, 16 January. O'Connor wrote, 'I now recommend the acceptance of Couston, Finlayson and Porritt's tender', and attached long and detailed conditions and a comparison of contract costs as against costs incurred within the department; the total Couston contract price amounted to £68,264. O'Connor ended by advising his minister that it would serve no useful purpose to call for tenders because the lowest tenderer 'might know little about it and the consequences, I fear, in that case would be very much deplored'.[3] This advice— in O'Connor's terms practical, logical and a means to greater efficiency—was nevertheless impolitic and provided the noose that his critics later tightened around his neck.

Although Rason, in his position as minister, could have made the decision himself without recourse to parliament, it is some indication of his uncertain grasp of ministerial office that he unwisely tabled Couston's offer and all the relevant papers in the Legislative Assembly on 21 January. On the following day, he moved that the pipes in connection with the Goldfields Pipeline Scheme be laid by means of a contract awarded to Couston, Finlayson and Porritt. The debate was adjourned until 26 January, three days after O'Connor had sailed for Adelaide.

That O'Connor should leave the State at this critical time had unfortunate consequences, but he could hardly be blamed for doing so. As Rason explained in parliament when O'Connor was being attacked in his absence, the South Australian Government had asked the Western Australian Government for a loan of the Chief's services in connection with the design of Adelaide's Outer Harbour:

> The South Australian Government had only recently lent us the services of one of their high officers and although it was recognised that it would be inconvenient for the State to part with [his] services…the Government could not but agree to the Engineer-in-Chief going.[4]

When it was generally known that the Chief had recommended switching to contract work on the pipeline, the suspicions harboured by his critics and which had lain dormant until then were suddenly aroused. The inference was that the pipeline work was in serious trouble and, further, that Couston's firm had been favoured without due process of calling tenders. The *Sunday Times* of 26 January trained its guns upon both these points in the first of its venomous broadsides. It described the minister, Rason, as 'a chameleon, the property of the highest bidder', and Couston's contract as 'a scheme that smacks of nauseous suspicion if not corruption from start to finish'. Then, singling out the Engineer-in-Chief, it accused O'Connor of fleeing the country 'to avoid responsibility for his work', describing him as 'that man whose whole career in the service of this State is fraught with sinister conundrums and peccant marvels'.

The debates in both Houses of Parliament were similarly undisciplined and wild with accusation and calumny. All the former prejudices of members averse to the scheme were released as if from a pressure chamber, and aired in a free-for-all. Inaccurate and defamatory claims were levelled against O'Connor and—echoing the *Sunday Times*—much inference was drawn from his absence in Adelaide. One of the most vocal critics was a querulous, diminutive Scot, Robert Hastie, whose general speech, it was written, 'was distorted with a particularly aggressive Scottish accent'.[5] Hastie accused O'Connor of over and over again giving assurances that the scheme would be completed on dates 'on which there was no prospect of completion'. He said that as far as he knew, not one of the public works under the control of the Engineer-in-Chief had been completed on time, and there was a sort of Freemasonry in O'Connor's department 'which prevented officers from expressing opinions'. Another member, Francis Wallace, former gold prospector and at that time storekeeper and shire chairman from the Lower Murchison district of Yalgoo, noted with sarcasm that O'Connor had been a student of Carruthers when in New

Zealand, and demanded to know whether 'further advice was being sought from that gentleman who had already advised them on what had turned out to be a successful failure'. Yet another critic, Doherty, said that the proposal for contract work 'took the cake for impertinence', and that it was a case for the Criminal Investigation Department. According to him, the affair was 'one of the greatest swindles that has ever been placed before Parliament'.[6]

The debate in the Legislative Council was no less defamatory. J. T. Glowery (South Province) attacked O'Connor's honesty and claimed that the contract would be a waste and extravagance in carrying out public work. The severest critic was aerated drinks manufacturer F. T. Crowder (East Province), who attacked the department, all its works and O'Connor equally. He had been proved wrong in his earlier opposition to Fremantle Harbour but apparently had learned nothing to soften his antipathy towards O'Connor and his projects.

In the Assembly, the Premier, George Leake, made only half-hearted attempts to control the spate of criticism. He, after all, had been an opponent of the scheme at its inception and was now in the uncomfortable position of having to defend it. 'I have never questioned [the scheme] from an engineering point of view but what has always troubled me is how this scheme should become a commercial success.'[7] His mood would not have been helped by an exchange of letters with Forrest, published in the *West Australian* of 24 January. Forrest, then Minister for Defence in the Federal Government and thus without any responsibility for State business, nevertheless wrote to Leake while in Perth on 14 January, expressing his regret that progress on the pipeline had been very slow. Although not in so many words, Forrest's implication was that O'Connor's work was being frustrated by government bureaucracy. Forrest justified his intervention by suggesting that the trans-continental railway—a project dear to his heart—was dependent upon a plentiful supply of water being provided along the first section of the route. To make matters worse, Forrest had shown the letter to O'Connor before sending it and O'Connor had advised amendments—a fact that later came to the attention of Leake himself.

The acrimonious debate on the Couston contract concluded on 3 February, when members criticized not only the contract but the Public Works Department and the conduct of the public service itself. Finally, a motion put by Charles Harper, Member for Beverley, was passed, 'that this matter be referred to a select committee with a view to getting as quickly as possible to the root of the difficulty'. Harper himself was appointed a member and voted chairman. Other members were W. J. George, J. L. Nanson, C. H. Rason and Henry Daglish. The group was hardly unprejudiced: all had spoken against the contract, and four of the five were known to have opposed the original Goldfields Pipeline Scheme.

The committee met over nine days, from 4 to 17 February, and examined Mephan Ferguson; Charles H. Hoskins; the co-inventor of the caulking machine, James Finlayson; T. C. Hodgson; and Couston himself—the latter called and re-called over four days. It has been pointed out elsewhere that the select committee, apart from consisting of prejudiced members, was conducting its inquiries without evidence given by the Engineer-in-Chief.[8] Rason, especially, was troubled by O'Connor's absence and sent a telegram to him suggesting that he get back as quickly as possible. The reply came from the Acting Premier of South Australia, saying that Mr O'Connor could not possibly get back any earlier: 'His leaving our work at this present stage would be a serious loss and vitiate all the work already done and seriously inconvenience my Government'.[9]

In the midst of the select committee's investigations, the *Sunday Times* of 9 February published its most scurrilous, defamatory attack on O'Connor, of such vehemence that it has since been widely accepted as significantly contributing to his death. With a black headline that left its readers with no room for doubt, it announced 'Corruption by Contract' and, underneath, 'Chummy Couston and O'Connor, the Pipe Track Scandal'. In two and a half columns of invective, wild accusations are hurled at the Engineer-in-Chief, describing him as 'Cunning and unscrupulous', and claiming that in all his undertakings he had 'misguided the State'—but worse, that O'Connor was corrupt:

> It is open rumour everywhere that this shire engineer from New Zealand has absolutely flourished on palm grease since the first day when the harbour works and the Coolgardie Water Scheme were agreed upon. If he is not now immensely rich there is some mystery somewhere. And apart from any distinct charge of corruption this man has exhibited such gross blundering or something worse, in his management of great public works that it is no exaggeration to say that he has robbed the taxpayer of this state of many millions of money…This crocodile imposter has been backed up in all his reckless extravagant juggling with public funds, in all his nefarious machinations behind the scenes by the kindred-souled editor of the *West Australian*. We need a court of justice in which to investigate O'Connor's relationship with the contractors.

We might wonder how newspapers at the turn of the century had the freedom to publish such wild and inaccurate libel unsupported by hard evidence, and we might also wonder what effect such writing had on the accused, and on the members of his family. O'Connor's daughter, the artist Kathleen O'Connor,

CORRUPTION BY CONTRACT

CHUMMY COUSTON AND O'CONNOR.

THE PIPE TRACK SCANDAL.

Toadies and Tricksters Who Should be Tried.

On Tuesday, the 27th of April, 1900, a banquet was given by his constituents to Mexican Morgans, at Coolgardie. The most prominent and remarkable incident on that occasion was the presence of C. Y. O'Connor, the Engineer-in-Chief for this State, to whom both the Premier Sir John Forrest and the Minister for Works took a back seat. As the *Kalgoorlie Sun* said at the time, "A still worse indecency was lic robbery adores him. From the beginning C. Y. O'Connor has been carted around by Ministers and made their spokesman— nay, the spokesman of the country, daring in his vaulting presumption to take Parliament itself to task. He did this at Coolgardie when he in the presence of the Premier and the responsible Minister for Works outlined the whole policy of the Forrest Ministry, whilst the Ministers present sat still with open mouths and staring eyes in the ecstacies of adoring admiration. Not

Portion of the *Sunday Times* article, 9 February 1902, thought to have contributed to O'Connor's death.

Courtesy Battye Library

was insistent in her opinion, expressed with bitterness towards the end of her life, that the *Sunday Times* drove her father to his death.[10] Whether this was true is open to question, but it was evident that when the Chief arrived back in Western Australia on 17 February, on the P&O ship *China*, and was confronted by the deteriorating situation, he was profoundly disturbed, a changed man. His companion on the voyage across the Bight was Charles Hoskins, with whom O'Connor had developed a close professional relationship. Hoskins reported that on the first day out of port, the Chief had shown his 'usual brilliance in conversation, a little like his old self'. On the second day, he spoke of Ireland 'and then broke off and without the least warning began talking about the weir and what had been said in the House as to its cost'. Hoskins went on to describe how, on the next day, in the midst of conversation, O'Connor 'broke off abruptly and said he did not feel well'.[11] O'Connor's youngest daughter, Bridget, met the ship in Fremantle and reported that her father 'was unusually depressed'.[12]

In the evening of 17 February, the day when O'Connor arrived back in Fremantle, the select committee presented its report to parliament. Its findings, published the following morning, did nothing to improve O'Connor's mood. Although Couston had been exonerated of any wrongdoing—'no evidence has been adduced…to justify the opinion that Mr Couston has acted otherwise than faithfully to the Government in his undertakings'—the judgment on Hodgson, O'Connor's second-in-command, was disquieting. The committee had closely investigated Hodgson's land dealings close to the site of the third pumping station at Cunderdin, laying him open to the charge that he had used his special knowledge of government work in the vicinity to his own pecuniary advantage. Although the committee stopped short of accusing him of nefarious practice, it concluded that his transactions in land near the pipe track amounted to somewhere approaching 40,000 acres, and that from his position in government service 'he might have taken it for granted that transactions of this magnitude would have acted to his prejudice in the mind of the general public'.[13] Equally disturbing for O'Connor was the conclusion of the report, recommending a further inquiry with additional powers: in other words, a Royal Commission with powers of subpoena, sworn evidence, and with the trappings of the formal courtroom.

On the 19th of the month, on the last afternoon of the session, Frederick Illingworth, the Colonial Secretary in the Legislative Assembly, successfully moved that a Royal Commission be appointed to 'Enquire into and Report upon the Conduct and Completion of the Coolgardie Water Scheme'. Rason, who might have been expected to support his departmental chief, instead supported the motion with Pilate-like self-justification:

> This work has caused the greatest anxiety to me as Minister, and when I was met with these proposals, as I was within a few days of taking office, I was convinced—I have no hesitation in saying it— that the working of the scheme in the past had been unsatisfactory.[14]

O'Connor must then have realized that it was he himself who would be the focus of the commission inquiry, that his management and competency would be on trial. He contemplated having to spend hours, perhaps days, being grilled by largely hostile inquisitors on matters of management policy and engineering practice that they would not properly understand. He, the Engineer-in-Chief, would be on public trial.

There is much evidence from this time forward that O'Connor became greatly distressed, that he was 'very much worried by all the attacks' and that 'he took nearly all as personal attacks on himself'.[15] In preparation for the Royal Commission, he instructed his secretary to go through *Hansard* and extract from it every allegation that had been made in connection with the Goldfields Pipeline Scheme, whether in favour of or against it. The secretary later confirmed how O'Connor was greatly disturbed: 'I saw him almost daily…he was upset and worried. I could always tell when he was worried'.[16]

The select committee had stopped short of recommending contract pipe-laying, and the work of the pipeline continued under existing conditions. Progress improved as the caulking teams became more familiar with Couston's equipment. Pumping stations numbers one, two and three were being erected, and representatives of James Simpson and Co. of London, manufacturers of the pumps, were on hand to supervise installation. Water pumping tests were planned to be made by the end of March. The most difficult terrain in the pipeline track over the Darling Range had been completed, and although not 'downhill all the way' in a literal sense, the work was becoming noticeably easier and faster. Railway trucks loaded with 28-foot long, 30-inch diameter pipes were getting nearer and nearer each day to Coolgardie, depositing their loads alongside the tracks. To guard against the extreme range of temperatures—frost

Unloading pipes from railway wagons beside the pipeline trenches.
Courtesy Battye Library 209351P

in winter and the heat of summer (113 degrees Fahrenheit or higher)—a special coating was devised for the pipes consisting of asphalt and coal tar applied hot and then sprinkled with sand. A simple sleeve method was used in joining one pipe to another, comprising a steel ring, 8 inches wide and slightly larger than the pipe itself, placed over the joint and then packed with lead.

Although O'Connor relied for much of the practical work of construction on Hodgson, his engineer-in-charge, to whom due credit must always be given, the final responsibility for and overseeing of this mammoth work was O'Connor's.

> Very few besides engineers credited the intense anxiety entailed during construction of a work of this kind, and none but engineers appreciated the difference between the reasons for success and failure.[17]

That O'Connor had this 'intense anxiety' for the project and at the same time was confronted with bitter personal attacks in the press and parliament, lack of official support, scepticism, and bureaucratic demands of antipathetic commissions of inquiry was inevitably to have an effect on his mental state. In our time, we would have no difficulty in recognizing grounds for a nervous breakdown: a clinical condition brought about by insupportable worry and an endemic tendency to mental depression. O'Connor's mental condition has been revealingly documented by his associates in those last days. Charles Hoskins met the Chief on business several times after returning from Adelaide and described how he appeared to be

> strung up to a terrible state of tension. He tried to multiply 43 by 3 and could not do it after trying several times…I could see from his personal demeanour that he was really in a terrible state of mind.

Hoskins saw O'Connor again a week later and found the Chief in much the same state of anxiety. O'Connor told Hoskins that he had not slept for two or three days, and while the two men were in the office an assistant came in with a little bottle of sleeping tablets. On yet another occasion, when visiting Hoskins at the engineering works, O'Connor 'could scarcely read the paper he was holding, his hand shook so much'. On Friday, 8 March, Hoskins saw O'Connor for the last time and on that occasion the Chief told him that

> the amount of abuse he was getting was almost more than he could bear…while we were talking he leant back [in his chair] and went to

sleep. It gave me quite a fright. There can be no doubt about it, the man's mind was gone.[18]

O'Connor seemed to have no one in a senior position to turn to. Forrest had gone. Martin E. Jull, the highly regarded Under Secretary for Works, a cooperative official as well as a deep personal friend who had been supportive in the past, was absent from the department.[19] Rason had shown himself to be, at best, uncommitted. The O'Connor family doctor, who might have helped, was away on holiday. In those last days, O'Connor visited his solicitor, which might suggest that he was contemplating taking legal action against his detractors. But Matthew Moss told him—surely with regrettable insensitivity—that 'it was stupid to pay any attention to what had been said in some of the papers'.[20] That O'Connor was ill and in urgent need of help must have been evident to those around him. That he did not receive help should not surprise us. Even today, in an age of social workers and counsellors, and when people are more open in expressing their feelings, there is still a shy reluctance to become involved in personal crises. At the turn of the century, a more formal age, even close friends would hesitate to cross the line between polite interest and giving intimate, compassionate help.

The weekend of 8–9 March registered the hottest spell then on record for that month. Heatwave conditions persisted for several days. Top temperatures ranged between 101 and 105 degrees Fahrenheit. The air was heavy and humid and exceptionally uncomfortable. On Saturday morning, 8 March, Hodgson and his assistants had witnessed the first test of 7 miles of pipeline in the vicinity of Chidlows Well, and it was claimed a success. One defective joint had been discovered but this was easily remedied. Hodgson advised his Chief of the results of the test and it was decided that they should both inspect the site the following Monday morning, leaving Perth by the 9.00 a.m. train.

The heat on Sunday morning was intense, the mercury registering 105 degrees in the shade. If O'Connor felt fit enough, he would have attended morning service at St John's Church with his family, as was his custom (he rented a family pew there). His depression was evident to at least one member of the family, his son Roderick, an apprentice engineer, who reported that 'in the evening he seemed worried and would not talk'.[21]

The following morning brought no relief from the heat. Even at an early hour, the low-pressure trough over the coast, which is so often trapped by high pressure in the Bight at that time of the year, produced still and stifling air, giving all who ventured out of doors a sensation similar to that of opening an oven door. O'Connor came out of 'Plympton House' at about 6.00 a.m. and made towards the stables. His groom, Arthur Lynch, saw him pass his bedroom window, and hastily went out to assist his revered employer. Lynch later described how O'Connor spoke to him, calling him by name—'Arthur, I want this horse early this morning. I want to catch an early train'—and then disappeared into the house again.

It was often the custom for O'Connor's daughter Bridget to accompany her father on his early morning rides, but on this occasion she was unwell and had decided not to go with him. Lynch described how O'Connor reappeared between 6.30 and 6.55 and took the reins of the horse, which was now saddled and ready; but instead of riding off immediately, he hesitated, and, tying the bridle to the kitchen door, he re-entered the house, saying that he was going to fetch something he had forgotten. Lynch did not see him come out of the house again but espied him riding out of the gate into Beach Street.[22] Lynch was probably the last to see him alive.

A mystery that we cannot solve is whether O'Connor had contemplated his next action for several days past and was merely awaiting a suitable opportunity, which now presented itself to him by the absence of his daughter, or whether it was a sudden decision that might not have occurred at all if he had had a companion that morning. The image of O'Connor on his horse preparing to ride away as usual, and then, suddenly, making his fateful decision to stop, dismount and re-enter the house to collect his son's revolver is one that will haunt Western Australian history forever. According to family tradition, Bridget, later Lady Lee-Steere, suffered much in the coming years on account of her absence from her father's side that fateful morning.

O'Connor's ride took him beyond Fremantle town and along the sands of South Beach to where the now disused Robbs Jetty entered the waters of Cockburn Sound. He pulled up there and, dismounting, let the horse go free. The beach hereabouts is a wide belt of soft sand, clean and white, hidden from the road by sparsely covered dunes. There must have been no one in sight. The water was flat calm at that time of the day, looking cool and inviting. It was here, on the day that O'Connor arrived in the colony eleven years before, that horseraces were first held. But O'Connor on that fateful Monday morning would not have been deterred from his purpose by the pristine beauty of his surroundings. His mind was at breaking point. Standing distractedly in 2 feet of water, he turned to face inland, towards the dunes and the rising sun in the east. For how many minutes he hesitated we do not know, but certainly he took the loaded revolver from his pocket and then, thoughtfully removing his dentures, placed these where the revolver had rested.

O'Connor's final act was to place the muzzle in his mouth and pull the trigger.

He fell dead at the water's edge in an instant, the noise of the explosion causing the horse to bolt into the sand dunes.

There were no witnesses. Only the evidence given at the inquest provides a reasonably accurate reconstruction of that terrible scene.

24

SHORTLY AFTER THE fatal shot was fired, Albert Cornwall, a labourer living a little south of Robbs Jetty in the vicinity of Owen Anchorage, started to make his way along the Rockingham Road towards Fremantle. His attention was attracted by an unusual sight. A horse with saddle and bridle hanging loosely was cropping the stunted, spiky grasses that thinly cover the low dunes separating the road from the sea in that area. Cornwall decided to investigate. There was no one in sight, apparently no owner of the horse. Puzzled, he went up to the animal and led it back along its own tracks, which made towards the beach. What he saw there shocked him: floating in the water close to the shore was the body of a man face up, about 16–20 feet from the jetty. The lower half of the torso was resting on the sand below the water, the face bobbing above the surface. Cornwall reported later that he could not identify who it was because of the blood, but he thought that it may have been Mr O'Connor 'by the clothes'. Much alarmed, he mounted the horse and made off towards the Fremantle police station to report what he had seen.

On duty that morning was Police Constable Richard Honner, who hurried off to Robbs Jetty at approximately 8.00 a.m. He was able immediately to identify the Chief, whose habit of taking a morning ride along the beach was well known to the police. He pulled the body out of the water. The skull, mouth and nostrils, he observed, were full of blood; the right hand rested on the chest and the left floated by the side. Foul play or suicide? Constable Honner made a search of the area and found a revolver in about 2 feet of water; it was loaded in four chambers and one of the bullets had been discharged. The riding whip was also in the water close by. No suspicious footprints were seen in the sand, he reported later, but there was evidence that the horse had bolted, frightened at the sound of the revolver shot. The body was then taken on a horse-drawn cart back to the mortuary at Fremantle, where it was formally identified by O'Connor's son, Roderick. An inquest was hastily convened in the Fremantle court at 2.15 p.m. and almost immediately adjourned until 10.00 a.m. the following morning, to allow for the results of an autopsy and the assembling of witnesses.

The news spread rapidly. The shock affected the whole community at all different levels. The Premier, George Leake, was in Melbourne, where he received the first telegram directed there by his deputy, Walter Kingsmill. It was sent before the full facts were known and spoke of O'Connor's death as 'apparently the result of falling from his horse'. Ironically, the expressions of universal shock, sorrow and effusive regret contrasted dismally with the denigration that O'Connor had so lately suffered. The Governor, Sir Arthur Lawley, was on holiday on Rottnest Island, and when he was informed he wrote prophetically to Kingsmill the following day:

> When I consider what a long and valuable service he had rendered in the development of Western Australia, with what zeal and patience he devoted to the carrying out of the great works, in the initiation of which he was instrumental and with which his name will always be associated long after they have been brought to successful issue, I realise how deplorable, from a rational point of view, is this tragedy.[1]

The Acting Premier replied to Lawley, pointing out that, in his opinion, the most pathetic aspect of the tragedy was his firm belief that the culmination of the two great works 'which will ever form monuments to the late Mr O'Connor's great and undoubted engineering ability' was now within measurable distance of completion, and that 'the mastermind which had hitherto guided these enterprises to a successful issue can never enjoy the consummation of the works'.[2]

Sir John Forrest was in Melbourne when he learned of his trusted Chief Engineer's death. His cable echoed Kingsmill's sentiments, speaking of his 'profound sorrow' and the

> gap which now cannot be filled…I mourn with the people of Western Australia the loss of one who has left behind a high and honourable record of splendid public service and I mourn the loss, also, of a dear and valued friend.[3]

Flags on government buildings were ordered to be flown at half-mast. Work was suspended on the harbour and the pipeline, and offices were closed in the afternoon of the following day 'so that those who wished to attend the funeral may do so'.[4]

When the inquest was resumed on Tuesday morning, 11 March, the first witness was Roderick O'Connor, who confirmed that the revolver belonged to him; it was one that he usually kept in a drawer in his room. He said he had found a document on his father's desk, in his father's handwriting, dated 10 March. It had all the appearance of being a suicide note written by a man under intense pressure:

> The position has become impossible. Anxious important work to do and three commissions of enquiry to attend to. We may not have done as well as possible in the past but we will necessarily be too hampered to do well in the immediate future. I feel that my brain is suffering, and I am in great fear of what effect all this worry may have upon me—I have lost control of my thoughts. The Coolgardie scheme is all right and I could finish it if I got a chance and protection from misrepresentation but there is no hope for that now and it's better that it should be given to some entirely new man to do who will be untrammelled by prior responsibility.[5]

And then O'Connor, ever mindful of his position as Chief Engineer even to the last, added his final instruction to his staff: 'Put the wingwalls to the Helena Weir at once'.

Five more witnesses gave evidence: Arthur Lynch, Police Constable Honner, Albert Cornwall, Charles Hoskins and the solicitor Matthew Moss. Their testimony of O'Connor's last few days has already been noted (see chapter 23). The coroner, Ernest Black, and three jurors returned a verdict of 'Death by his own hand through a bullet wound from a revolver at Robbs Jetty while in a state of mental derangement caused through worry and overwork'.[6]

An understated heading in the *West Australian* described the funeral on the afternoon of 11 March as an 'imposing spectacle', but so large was the crowd along the route, and so well attended by vice-regal representatives, the Acting Premier and other government ministers, officials and leading citizens, that it could more accurately have been described as a 'state occasion'. The cortège wound its way along Parry Street and into High Street in an easterly direction towards Fremantle cemetery. Pallbearers and chief mourners included Kingsmill; O'Connor's minister, C. H. Rason; Septimus Burt; O'Connor's leading engineering colleagues—Hodgson, E. E. Salter and A. Dillon Bell; and members of the family—his sons-in-law, George Julius and C. Y. Simpson, and his sons,

Exhibit.

The position has become impossible

Anxious important work to do and three commissions of enquiry to attend to

If we have not neglected our business and the past we

We may not have done as well as possible in the past but we will necessarily be too hampered to do well in the immediate future

I feel that my brain is suffering and I am in great fear of what effect all this worry may have upon me — I have lost control of my thoughts

The Coolgardie scheme is all right and I could finish it if I got a chance and protection from misrepresentation but there is no hope for that now and its better that it should be given to some entirely new man to do who will be untrammelled by prior responsibility

10/3/02

Put the wingwalls to Helena Weir at once —

O'Connor's suicide note, written the day he died.

Original held at SROWA

Frank, Roderick and Murtagh. The long, sad procession, which extended nearly a mile in length, was led by 150 employees of the harbour works and the Goldfields Pipeline Scheme on foot, closely followed by a long line of carriages and then an estimated 'thousand gentlemen'. Several thousand members of the general public lined the route. At the graveside, Archdeacon Watkins, the O'Connors' long-time friend and incumbent of St John's, Fremantle, conducted the service at the graveside. On the jarrah coffin was a plate bearing a simple inscription—Charles Yelverton O'Connor, died 10 March 1902, aged 59 years—and below it, an embossed anchor and a cross bearing one word: 'Hope'. There were so many wreaths that they covered the hearse, and completely filled a second carriage.[7]

In the days following the funeral, generous tributes continued to pour into the State, not only from within Australia but from overseas. A message of condolence was received from London from the previous Governor and his wife, Sir Gerard and Lady Smith, who would have counted O'Connor as one of their friends. New Zealand's Prime Minister, Richard Seddon, expressed in a telegram his deepest regret at the death 'of my old friend' and sympathized with 'the irreparable loss to your colony'. Leader of the Opposition in the New Zealand Parliament, William Rolleston, who had known O'Connor well, wrote:

> He has undoubtedly been one of the most remarkable and useful pioneer settlers…he combined gentleness and amiability with force and vigour of character and intellectual activity to an extraordinary degree.[8]

The New Zealand papers carried the news of O'Connor's death, linking it with his worries over the pipeline. The *Examiner* described O'Connor as 'a man in 10,000…hard work had no fears for him. And the loss of his labour was a loss to the colony'.[9] *The Engineer*, the profession's leading magazine, published in London, devoted two columns to an obituary and summarized with the statement 'no engineer has ever done more than he for an individual colony'.[10]

These and many similar tributes are in startling contrast with the cruel accusations that had been published before O'Connor's death. This posthumous judgment on him—prompted perhaps to some extent by feelings of remorse— was directly contrary to the opinion of Mark Antony when musing on the death of Julius Caesar. It was the good in O'Connor that lived after him; the evil (if any) was buried with his bones.

Self-justification, however, and not remorse, was the response in the *Sunday Times* the week following his death. That newspaper's report opened by

O'Connor's grave and memorial in Fremantle cemetery, erected from subscription
by Public Works Department staff.

Photograph by the author

expressing sympathy for those bereaved and then explained that 'all our criticisms of the deceased gentleman have been sans malice and purely devoted to the public welfare'. The writer continued disingenuously, and with a curious inversion of the facts, accusing those who had blamed the paper for O'Connor's death of slandering the departed. 'No honorable, strong minded man is afraid of the severest public criticism'—implying, surely, that because O'Connor was clearly affected by public criticism he may have been dishonourable. The case against him was then continued in more subtle, ambiguous language:

> It is only men of weak intellect, or men who feel and realise some oppressive guilt upon their consciences who are perturbed by ques-tions upon their conduct...At this moment we even prefer to say nothing in justification of our past attitude towards the man, when living, beyond saying that towards him, as towards all others, we have always been actuated by the highest sense of public duty, which we cannot regret even in the presence of death itself. We are content to wait for the vindication of Time.[11]

The vindication of Time was not long in arriving to prove that at least one of the *Sunday Times*'s accusations was entirely false. Instead of being 'corrupt' and of having 'flourished on palm grease', O'Connor died a relatively poor man. He owned two horses valued at £20 and £12 each, and a cow (which, according to Kathleen, seldom gave any milk), £15. The value of his furniture, including a piano, amounted to £160; his personal effects, clothing and books, £35. The salary owing to him up until 9 March was £36 5s 10d. Among his liabilities were listed a small overdraft at the Union Bank of Australia; municipal rates and rent due on 'Plympton House'; various grocers' bills; an account with the local blacksmith; and accounts owing to drapers and general stores. When the liabilities were subtracted from the assets, the credit balance amounted to £189 5s 10d.[12] The Chief owned no property and thus his family would have been destitute had it not been for his modest life insurance.

In September 1902, six months after O'Connor's death, the government passed an Act to provide an annuity for his widow, amounting to £250 per annum. Although some voices were raised in objection—it being argued that O'Connor had been one of the highest paid civil servants and should have made adequate provision for his family himself, and that providing a pension in this case might create a precedent—the main speeches in the second reading were supportive, some amounting to eulogies, a public acknowledgment of O'Connor's selfless energies on behalf of the State. His minister, C. H. Rason, who had a reputation for effective, eloquent speaking, began by agreeing that O'Connor's case was an exception and should not be taken as a precedent, but he then went on to praise O'Connor and his eleven years of work in the highest terms:

> Eleven years of probably the hardest work any man in his position was ever called upon to perform. They were eleven years of gigantic undertakings, works of very large magnitude, which were dependent mainly on his engineering skill for their success. But they were dependent not only on his skill, but on his application to the discharge of his duties.[13]

Rason then pointed out that O'Connor had taken no leave and that the money owing to him for leave and other entitlements would amount to a sum almost sufficient to provide an annuity for his widow.

Among those who spoke in support of the Bill was the Member for Claremont, J. C. Foulkes, who said that he had counted himself a friend of O'Connor, and described him as setting an admirable pattern to all holding high position. 'It was his constant pride that he kept himself clean-handed and religiously abstained from making any investments whatsoever in this state.'[14] (If this observation is from personal knowledge, and not mere hyperbole, we may have the explanation of why O'Connor never bought, sold or owned property, as others in his social position so readily did.)

If the family followed the course of the debate, they must have been relieved when the Bill was passed, but also embarrassed that their affairs had been publicly debated. Foulkes had described how he had seen Mrs O'Connor travelling in a second-class carriage and had felt ashamed that she, the widow of the man who had reformed and constructed the railways—'a memorial to his memory'—was not even the recipient of a free pass. The family might also have wondered at the inconstancy of a parliament that had once been the scene of so much vilification of O'Connor and now extolled his virtues and lauded him as a hero.

Four of the children were still living at home: Roderick, aged 20, Kathleen, 25, Bridget, 18, and Murtagh, 13. Roderick assumed the responsibilities of management; his account book shows the extent to which the monthly payment of £20 16s 8d from the annuity had to be stretched to cover living expenses.[15]

How much the family suffered at this time can only be imagined; there is little documentary evidence. Added to the children's overwhelming loss of a beloved and respected father would have been the unspoken sense of shame that was attached to suicide one hundred years ago. In Christian ethics, it was regarded as not only wrong but illegal. Up until the early 1950s, a failed suicide attempt could lead to prosecution. Kathleen, as already noted, expressed her sorrow by anger directed at her father's critics. She kept a scrapbook of cuttings and photographs relating to her father's work, and carefully underlined significant or damaging criticisms. In her reminiscences of her father, she makes a brief reference to the lack of support given him: 'If only he had had the right support, the whole world would have been a joy to him…for a man with a highly sensitive nature it is rather frustrating'.[16] Bridget's views seemed less restrained when, years later, as the first Lady Lee-Steere, she described her father in a broadcast talk:

> [It was] the continued criticism of untrained minds, petty jealousies in party politics, and a sad lack of Government support, which eventually weighed down and broke the fine and brave spirit of an overworked, highly sensitive and conscientious man. We lost a wonderful and devoted father, and a great joy and inspiration out of our lives.[17]

That joy, inspiration and affection are hinted at in Lady Lee-Steere's last letter to her father, addressed to him at the Adelaide Club, dated 6 February. Signed 'Biddy', as she was known then, she thanks him for his letter to her and, after exchanging family news, goes on to describe how his horse Moonlight had been ill:

> He must have been very bad for he lay down all the time for a day or two but he is quite well again now. I have not seen Aileen and the baby yet but I think Eva and I are going up to see them on Saturday. This is a very short letter but if it does not catch this mail it might not get there before you start for home. With best love from all of us. I am ever your loving daughter, Biddy.[18]

In spite of Biddy's efforts, the letter did not reach her father before he left to return home on the *China*. It was redirected, returned to sender via the Weld Club in Perth, and arrived at 'Plympton House', unopened, unread, after her father's death.

The Royal Commission appointed to 'Enquire into and report upon the conduct and completion of the Coolgardie Water Scheme' interviewed its first witnesses on Wednesday and Thursday, 5–6 March, a few days before O'Connor died. William Kirk, contract and stores clerk, and George Kitchen, assistant engineer, gave evidence concerning the inspection and acceptance of the pipes and other equipment. O'Connor would surely have read the verbatim report of the opening proceedings in the daily paper; how Kitchen had rejected 212 pipes at the factory because they were not properly cleaned and how there had been a dispute with the manufacturer. Although there was nothing particularly incriminating in either of the men's evidence, it would have been clear to the Chief that the mood of the commissioners was hostile and that they were bent on finding fault. The likelihood of the inquiry dragging on for several months, involving minute inspection of documents and memoranda, the hearing of claims and counter-claims by disgruntled employees and contractors, the citing of poorly understood technical reports, and belligerent questioning, all must have amounted to an appalling prospect for O'Connor and added to his mood of despair. At the end of the second session on the Thursday preceding his death, the commissioners adjourned until the following Tuesday, 11 March, at 2.15p.m. (In the event, the third session was postponed; the members were not to know that they would be attending O'Connor's funeral at that time.)

Stockpile of pipes ready for delivery at Mephan Ferguson's works at Falkirk
(now known as Maylands).

Courtesy Battye Library 120932P

The commissioners held forty-seven meetings and interviewed fifty-nine witnesses. They produced two interim reports, the first on 2 April recommending that

> it is desirable for the Government to invite tenders for those portions
> of the work remaining to be done east of Cunderdin, namely the
> pipe-laying, jointing, caulking and covering and such other work as
> may be suitable for public tender.[19]

They thus showed that they endorsed the principle of contract work originally recommended by O'Connor, but insisted on calling for tenders. Such a cumbersome and disruptive process against O'Connor's advice was quietly ignored, largely because the competency of the men working the caulking machines had so improved by then that satisfactory progress was being made.

The brunt of the inquiry was borne by Hodgson who, as engineer-in-charge, was said by the commissioners to be 'practically in supreme control'. They stated that there was 'not recorded any decision arrived at by the Engineer-in-Chief without Mr Hodgson's advice thereon having been taken'. The second interim report, published on 23 May, was highly critical of Hodgson, the evidence against him centring on his acquisition of land in the vicinity of the pipeline and pumping station at Cunderdin, his borrowing money from Couston, the favoured contractor, at advantageous interest, and what the commissioners maintained was his self-serving advice, which had rejected an engineering requirement to relocate the Cunderdin pumping station at Tammin. As a result of these allegations, Hodgson was suspended from duty pending a disciplinary inquiry, which in the event never took place.

Based exclusively on the findings of the Royal Commission, T. C. Hodgson is revealed as the villain in the drama, a man who corruptly used inside information for his own gain and manipulated the course of the pipeline to his own advantage. Further, the commissioners found that 'it was the discovery by Mr O'Connor of the degree to which his implicit trust had been misplaced, and the financial results proceeding therefrom, that finally unbalanced an already overstrained mind'.[20]

There is, however, a case to be made in defence of Hodgson, an argument that he was not as seriously implicated as the Royal Commission made out. The argument runs that the worst that can be said of him is that he was indiscreet in borrowing money from the contractors, Couston and Finlayson. And as to his land dealings, the first 100 acres were purchased on 9 June 1896, and the second on 9 July in the same year, both dates preceding the decision of the London

Commissioners to endorse Cunderdin as the site of pumping station number three. The accusation that, subsequent to the London Commissioners' report, he falsified the best engineering advice favouring relocation to Tammin is an extremely technical one resting on the use of the locking-bar pipes: it was said that these would reduce friction and improve head or pressure to such an extent as to allow repositioning of the third pumping station. This claim, however, has been contested by later hydraulics engineers, who argue that the original location of Cunderdin was proved later to be the most satisfactory site, and both O'Connor and Hodgson had believed it was at the time.[21] And this conclusion has certainly been borne out in modern times as extensions and branch pipelines have been added to the main trunk. 'Future generations were to benefit from the judgement of O'Connor and Hodgson and to reflect that engineering was more a matter of judgement than an act of calculation.'[22]

The commissioners carefully avoided any critical reference to O'Connor, such was the feeling of sorrow in the community at his untimely death: *de mortuis nil nisi bonum* (let nothing but good be said of the dead). They did, however, allow themselves to mention that 'his deplorable death has necessarily hampered the Commission in its work', and that had it been possible to examine him, 'a greater degree of light would have been thrown upon the work and his relations to it than at present is available'.[23]

Hodgson resigned in August 1902 and lived on as a much respected and successful landholder near Cunderdin, running horses and cattle as well as growing fresh fruit. J. S. Battye's historic 1912–13 *Cyclopaedia of Western Australia*, apparently unaware of the irony, described how 'his special experience in water conservation served him in good stead in his private enterprise'.[24] He was voted president of the Cunderdin Farmers and Settlers Association, and died at the age of 72 in 1935. W. C. Reynoldson, who took over as engineer-in-charge after Hodgson left, wrote a letter to the *West Australian*, dated 11 December 1935, following Hodgson's death, listing his important role in the successful construction of the pipeline and concluding, 'My object in bringing this matter before the public is to give credit where I know credit to be due'.

The Minister for Works, C. H. Rason, while stating—surely precipitately—that Hodgson's 'reputation as an engineer…has gone forever', went on to acknowledge his major contribution to the success of the scheme:

> I think it is only right to say that many good things that have been done in connection with this scheme are distinctly due to the experience and ability of Mr Hodgson; and this scheme undoubtedly owes to Mr Hodgson a very great deal.[25]

Throughout the remaining months of 1902, work progressed quietly and efficiently on the pipeline, under the guidance of the new engineer-in-charge, Reynoldson, and the Engineer-in-Chief, Charles Palmer. Over 90 miles of piping had been laid prior to O'Connor's death; the remaining 260 miles were completed by early 1903. The whole period of construction, including the building of Mundaring Weir, had been less than five years—a remarkable achievement given the early difficulties, and the necessity of importing all metal for pipes, the cement, the valves, the lead for jointing, the pumping machinery and ironwork, and much other material. The total cost, including all extras, contingencies and establishment charges, was £2,660,000, an excess of £225,000 on O'Connor's original estimate. 'This can hardly be considered a large excess when it is remembered that the original estimate was based on tentative data prior to survey.'[26]

The grand ceremonial opening of the Goldfields Pipeline Scheme took place on Saturday, 24 January 1903, a little over ten months after the architect of the scheme had died. On the previous Thursday, Lady Forrest had ceremoniously started the pumping machinery at pumping station number one below the Mundaring Dam wall, although the water had been flowing through

Much of the original Goldfields Pipeline was laid in trenches to protect it from intense heat and winter frosts. Here an exposed section gives a graphic picture of the grandeur of C. Y. O'Connor's great aqueduct.

Courtesy Battye Library 783P

[239]

the pipes well before that date. Two days later, a fleet of special trains conveying federal and State guests from the coast arrived in Coolgardie before 8.00 a.m. and were met by the mayor, Mr A. P. Wymond, and conveyed to the United Club Hotel for breakfast. The town was gaily decorated with flags and bunting. Prior to the official ceremony, timed for 11 o'clock, a colourful procession, which included two bands, pipers, camels and the fire brigade, wended its way through the town to the Exhibition Building, where children sang the National Anthem. Sir John Forrest, with his 'amiable wife', was 'accorded a splendid reception'.[27] Like royalty, acknowledging the cheering crowd, they rode in an open phaeton, drawn not by horses but by twenty-four members of the Boy Sailors' Brigade dressed in their sailors' uniforms. At the exhibition grounds, a dais had been prepared alongside the taps and pipes, which would supply the town with fresh water. Forrest was given a rousing welcome, recognized, as he was, as the instigator of the scheme, one who had from the start supported it in parliament and fought for it. The people were in a mood to acknowledge him as their saviour. It was left to Forrest to remember the architect of the scheme, which he did in a speech expressing his sadness that the great engineer 'had not lived to receive the honour so justly due to him'.[28] After the speeches, Sir John turned the guiding wheels of the valves and a clear stream of water spurted into the air from a fountain situated a few yards from the platform. The band played 'Rule Britannia' and then the guests sat down to a lavish luncheon before moving off in the trains for the second stage of the ceremonies at Kalgoorlie.

As colourful, spirited and acclaimed as the Kalgoorlie ceremonies were, they were not devoid of misfortune. For a start, the heat, even at 5.00 p.m., was intense—113 degrees Fahrenheit in the shade—and the temperature at the opening ceremony at Mt Charlotte Reservoir, in the open air under a blazing sun, was considerably higher. The report of the event in the *West Australian* suggested, tongue in cheek, that the great discomfort suffered by the crowd from the heat would actually serve to show just how valuable the supply of fresh cool water was in the circumstances. If the weather had been cool, the water may not have been so avidly awaited. The same paper grumbled that the arrangements for the large contingent of visitors was 'lamentably inadequate'. Hotel accommodation was not available for all of them and many had to 'content themselves with sleeping quarters in the trains which had brought them from the coast'. And, unlike the arrangements at Coolgardie, little provision was made for conveying the guests from the hotels to Mt Charlotte; faced with a long walk in the heat, 'several declined to make the journey'.[29]

But these setbacks did not seriously mar the grandeur of the occasion. It was written that 'Enthusiasm was the predominant feature' and the proceedings were carried out 'on a scale of magnificence'. All agreed that it was a red-letter day—a 'golden day'—for the goldfields.[30]

Later that evening, after the water had started flowing into the giant Mt Charlotte Reservoir, all the official guests, federal and State, were entertained at a banquet in the car barn of the Electric Tramway Company. The Mayor of Kalgoorlie, Mr N. Keenan, caused embarrassment in his speech of welcome when he suggested that the mines would buy the water because the scheme was a State undertaking, and not because the water would be cheaper than if procured from other sources. Forrest, in his reply, immediately repudiated this view, and censured the mayor for his 'unbusinesslike proposition'.[31]

Among the many official guests at the banquet were C. Y. O'Connor's eldest son, George Francis (Frank), then a qualified engineer like his father, and his widow, Mrs Susan O'Connor. For these two, at least, the shadow of the architect of the scheme, the man who made the ceremonies on that day possible, must have been palpably present; his spirit must have permeated the proceedings. But as far as the euphoric crowds were concerned, O'Connor could not have been there; he had been dead for the last ten months. And yet for future generations his very absence on that occasion seems to haunt us like a spectre in a historic photograph, or perhaps a blank space where he should have appeared in the picture. We look for him next to his friend and patron Sir John Forrest, where he was so often to be seen in the past at the numerous ceremonies that marked stages in the progress of his works. His absence at Kalgoorlie on that day—that blank space—is an uncomfortable reminder of the cruelty of circumstances, the fickleness of public opinion, and our helplessness in the face of others' personal sufferings.

> About suffering they were never wrong,
> The Old Masters: how well they understood
> Its human position; how it takes place
> While someone else is eating or opening a window
> or just walking dully along…[32]

Official opening of the Goldfields Pipeline Scheme at Coolgardie, 24 January 1903.
We look for the familiar image of 'the Chief' but he is not there.

Courtesy Battye Library 2570B (5054P)

Epilogue

ONE HUNDRED YEARS after C. Y. O'Connor's death, his pre-eminent status in the history of Western Australia is assured: he has become a legendary figure, revered as one of the progenitors of the modern State, and instrumental in laying a foundation for its subsequent industrial, mining, agricultural and commercial achievements. His two major works, Fremantle Harbour and the Goldfields Pipeline—aided, it must be said, by his tragic death—have spread his fame far beyond State boundaries; he is now generally recognized as a national figure. In Western Australia, a federal parliamentary electorate and a suburb are named after him, and there are at last count four O'Connor streets in the metropolitan area. He has a popular museum dedicated to him in the hills close to Perth, and the National Trust has launched the Golden Pipeline Project designed to preserve O'Connor's original pumping stations and pipeline as a national monument and Heritage Trail. As well as Porcelli's statue in front of the Fremantle Port Authority building, there is a bronze bust overlooking Mundaring Weir, and yet a third statue by the sculptor Tony Jones, symbolizing O'Connor's last ride and death, has been placed in the sea close to the site of Robbs Jetty south of Fremantle. And, yes, there is even a pub named in his honour in West Perth.

Nor does O'Connor belong to Australia alone. Although less well known in New Zealand, in spite of his twenty-six years there, and although he did not undertake a project of the magnitude and glamour of the pipeline there, he was without doubt a key figure in that country's early development. Among New Zealand civil engineers—that sadly anonymous profession—he ranks with the highest.

It is interesting to speculate on what future O'Connor might have enjoyed had he not ended his life when he did. He surely would have survived the Royal Commission, even if some small blame were to have been attached to his administration and his supervision of staff. The rapidly approaching and successful completion of the pipeline and Fremantle Harbour would surely have overridden any petty criticisms. He would have taken his place next to Sir John Forrest at the opening ceremony in Kalgoorlie when the water reached there on 24 January 1903, and there he would have been received with much honour and

Bronze bust of C. Y. O'Connor overlooking Mundaring Weir in the
Darling Range east of Perth.

Photograph by the author

praise. Within a few years of retirement, his CMG would have been elevated to a KCMG for his services to the State, and perhaps he would have lived at ease and in comparative obscurity in Claremont or in a modest colonial house in Fremantle. As we have seen, he was not a wealthy man and it is difficult to imagine him, under those circumstances, owning property. Shortly after he arrived in Western Australia, he corresponded with an Albany estate agent about the purchase of a small country ranch for his horses, but those early negotiations came to nothing.

Of course, the above is mere hypothesis: we know that O'Connor chose to end his life on the very eve of the completion of his two major works. And without wishing to diminish in any way his stature and his undoubted genius, his dramatic suicide has done much to establish him in the public mind as a folk hero: a tragic figure in the tradition of Greek drama. That C. Y. O'Connor killed himself is a fact of history. Why he killed himself is not so readily explained.

All those who admire O'Connor and have studied him, and all those who have written about him, have pondered his death. We wish it not to have happened in the way it did. Like watching *Othello*, or any other great stage tragedy, even though we may know the plot, for the duration of the performance we desperately hope that the death in the last act might not happen. So it is with O'Connor. The question has been asked over and over again: why did O'Connor die in the gruesome way he chose? Why did he take his own life on the eve of celebrating his triumphs, and leave his body in the sea, his children bereft of their father, and his widow mourning him in reduced circumstances?

As befits the death of a legendary hero, there have been various theories postulated. One of the longest enduring and most widely believed—even taught in schools—was that the water did not arrive at its destination when O'Connor had predicted it would (owing to a supposed miscalculation). As a result he despaired, believed his scheme a failure, and took his own life.

Another improbable theory is that he did not commit suicide at all but was murdered by his enemies. Who those enemies were and what they could hope to gain from his death are never explained.

A more recent theory has been mooted that a curse was put on him by the Aboriginal people of the Bibbulmun tribe, whose ancient sacred site was destroyed by the blasting of the rock bar at the entrance to the Swan River.[1]

Yet another theory has him guilty of the misdemeanours he was accused of in the newspaper and, rather than risk exposure and shame following the Royal Commission, taking the only safe way out.

One of the more responsible and closely argued theories of recent years was advanced by Emeritus Professor Martyn Webb in a paper read to the Royal

Western Australian Historical Society, entitled 'The Death of a Hero'. Webb argues persuasively that O'Connor's lieutenant, T. C. Hodgson, was the real villain in the drama, being engaged in corrupt land deals and giving specious advice on the siting of the Cunderdin pumping station, much to his own advantage; when O'Connor discovered this at a late stage, he realized that he would be held to account and would have to accept full responsibility as chief executive of the Public Works Department. As Webb says:

> O'Connor must have become increasingly worried and even more distressed, and possibly outraged, as, step by step, he began at long last to discover what had been going on in his name, beneath his very nose, and under his signature since as long ago as June 1896 when, as the newly appointed engineer-in-charge, Thomas Hodgson had begun purchasing land around Cunderdin.[2]

Resulting directly from this, O'Connor, being a man with a highly developed sense of honour, was faced with only one course open to him: like Brutus, who 'set honour in one eye and death in the other, and looked on both indifferently', he fell on his sword.

To us, living at a time when legislators and executives in public companies who are found guilty of lying or corrupt practices retire with generous super-annuation and 'golden handshakes', suicide seems a wantonly unnecessary solution. But a century ago, notwithstanding the illegality, it was not uncommon among certain classes of professional people—and was even considered coura-geous and expiating—to 'take the honourable way out'.

As persuasive as Webb's theories of O'Connor's suicide may be, they remain conjectural, at best not proven, and raise as many questions as they answer. They are based on the assumption that Hodgson was as corrupt as the Royal Com-mission found him to be and that the commissioners were fair and unprejudiced. And yet we know that every member of the commission had been critical of the Goldfields Pipeline Scheme at one time or another and had spoken against it in parliament:

> The Commissioners appear to have begun their task with the assump-tion that everything touching the scheme was wrong and that those implementing it were culpable of misdemeanour...Most damaging of all, officers of the Coolgardie Water Supply Branch must have felt themselves open to attack. That attack directed in full force against Hodgson and Couston.[3]

Hodgson's farming activities, and his position as an influential citizen in Cunderdin, had been overt and known to the world at large for several years. He had been voted chairman of the town's Progress Committee in May 1901, and signed his name to a petition to the government urging increased railway facilities. Reports of the committee meetings and Hodgson's association had been printed in the daily newspapers for all to read.[4] It is difficult to accept that O'Connor would not have known about Hodgson's landholdings in Cunderdin, as others in the Works Department must have done. All politicians—most notably the Forrests and John Forrest's Cabinet ministers—merchants and professional men and anyone who could save sufficient money in those days bought and sold land; O'Connor, in not doing so, was an exception. Land was where fortunes were made. To invest in land and profit by it was utterly respectable.

It is perhaps significant that although the commissioners were highly critical of Hodgson and as a result he was suspended from duty in May 1902, he was never charged in law with any offence. He lived in retirement on his property in Cunderdin, a prominent citizen, until his death in 1935.

According to Public Works Department staff, Hodgson was interviewed by his Chief shortly after O'Connor's return from Adelaide. We do not know what passed between them, but the same staff reported that Hodgson offered his resignation there and then, but O'Connor would not accept it.[5] This begs the question: why would a man of such fearless probity as O'Connor, if he were convinced of Hodgson's wrongdoing, not have accepted Hodgson's resignation there and then? Surely the case would seem to have demanded it. O'Connor was, according to Webb's argument, 'possibly outraged' when he discovered what had been going on his name. But there is no hint of outrage, and on the evidence of his staff it seems that O'Connor was not sufficiently convinced of Hodgson's guilt. And if he was not convinced of his wrongdoing, where does this leave Webb's theory that O'Connor took his life largely because he felt responsible for Hodgson's nefarious dealings? Of course, it might be argued in turn that, at that interview, Hodgson could have said to O'Connor, 'You knew all about my land dealings, and so if I am guilty, then so are you'. But this is highly speculative and, as Webb writes, O'Connor was only just then discovering what had been going on.

In another part of 'The Death of a Hero', Webb writes that 'O'Connor reached his decision to kill himself by rational means and as part of his honour system, and as a reflection of his hero status'. Evidence suggests otherwise. Eyewitness accounts of O'Connor's last few days, descriptions of his behaviour at that time, show that he was far from acting rationally; that he was, in fact, 'out of his mind'. His distressing symptoms had been observed on board the ship from Adelaide before he had read the parliamentary report implicating Hodgson. We know that he was suffering from insomnia and from headaches, and yet he dozed off at his desk in the middle of a conversation. He, an experienced graduate surveyor, was suddenly incapable of doing the simplest multiplication; his hands shook and he could not read a document. He could not concentrate and he himself expressed the fear that he was losing control of his mind. We have seen evidence that he would abruptly change subject in conversation. All these, rather than showing rational behaviour, demonstrate that he acted very irrationally in the days leading up to his death. The symptoms described by eyewitnesses in detail during those last days indicate, more likely, that O'Connor was in the grip of acute depression—a state that was possibly endemic and largely hidden from the world, building up over a long period. In addition, his intense anxiety over the Goldfields Pipeline Scheme, the delays over which he had little control, the lack of support from government, the cruel and unfounded newspaper criticisms, and, it must admitted, the new discoveries of perceived staff irregularities—all these factors would have compounded to reach a climax known to modern psychologists as 'acute anxiety disorder'. As one leading authority states, 'Suicide is not caused by a single factor, but rather a complex combination of factors'.[6]

The autopsy performed after O'Connor's death revealed that he had contracted cirrhosis of the liver, which points to a heavy consumption of alcohol over a long period. One of his scribbles in a Western Australian diary reminds him to collect a bottle of brandy 'from the club' before catching his train—not unusual perhaps for a long journey in those days but, in view of the cirrhosis, of more than usual significance. There is therefore some evidence to suggest that O'Connor would have sought relief from the exceptional pressures of his work by drinking heavily, especially towards the end. Descriptions of his behaviour and his appearance tend to support this view. O'Connor, we know, was a very private man; although outwardly courteous and friendly, he evidently kept much to himself, revealing little of his inner thoughts. Normally calm in a crisis, he would occasionally erupt in a temper, according to his daughter Kathleen. But these outbursts appear to have been quickly suppressed. This is consistent with a view that O'Connor may have been suffering bipolar manic depression—a condition characterized by periods of increased energy and heightened impulsiveness alternating with dysphoric moods and depressed cognition.

Towards the end, when the crisis came, it is clear that O'Connor needed help, but perhaps he felt there was nobody to turn to; and even if there had been someone, he was probably convinced that he could not communicate his fears and his anxiety. It is often this feeling of total isolation of the individual that leads to a suicide attempt. In order to survive periods of great stress, the body requires periods of calm and distraction to regain balance so that clear thinking can be restored. That O'Connor did not have that necessary period of restoration towards the end is clearly evident.

O'Connor's life was both a triumph and a tragedy. He pulled himself up from disadvantaged beginnings in a famine-torn Ireland to become a master of his profession, inspired and inspiring to others. By his own energy, probity, vision and application, it could be written of him at his death that 'no engineer [had] done more than he for an individual colony'.[7] His first colony, New Zealand, treated him shabbily towards the end of his stay, and when he arrived in Western Australia he fell victim to the notorious 'tall poppy syndrome', to be cut down by jealousies and misunderstandings. Western Australia at that time, and even later, was an oligarchy, close-knit, insular and generally suspicious of outsiders, resentful of the newcomer, especially those who showed exceptional talent and lofty self-confidence. This attitude has continued until quite recently and is not confined to one State alone. An example in our own time, providing a curious if tentative parallel with O'Connor, is the brilliant Danish architect Joern Utzon, creator of the Sydney Opera House. At first he was welcomed, as O'Connor was, and supported by both government and community, but at a later stage, when difficulties arose, support was gradually withdrawn and his professional reputation undermined. Now, thirty-five years later, Utzon is forgiven and revered—and has even been asked to return. But, like O'Connor, although for a different reason, he cannot do so, and we are left with regrets and the loss of someone who could have contributed so much more. Burley Griffin is another example of a visionary who was not appreciated while he was working in Australia and left to continue his work elsewhere. There are many other illustrious names, especially among artists and scientists, who have become exiles.

With the arrival of multiculturalism in Australia, a new climate of tolerance and appreciation has become well established. We now recognize the rich contribution made by doctors and scientists, restaurateurs and technicians, bankers and industrialists from other countries, and sometimes we honour them with decorations.

In the final lines of his play *St Joan*, Bernard Shaw makes the spirit of the eponymous young heroine, lately burnt at the stake by the French (with the connivance of the British), cry out: 'Oh God that madest this beautiful earth, when will it be ready to receive Thy saints? How long oh Lord, how long?'. It may not be too fanciful to apply the same question to Australia and ask whether this beautiful country is now, at last, ready to recognize and support its gifted immigrants while they are still alive instead of erecting monuments to them after they have departed.

More than 100 years after O'Connor's death, it seems that St Joan's prayer has been answered; the time of acceptance has arrived for Australia. However, it has arrived too late for O'Connor and others.

In one sense, the artist, the craftsperson, the architect and the engineer do not die, but live on in the works they bequeath to us. The Latin inscription in St Paul's Cathedral, written of the architect Sir Christopher Wren, might well be appropriate for O'Connor also: *Si monumentum requiris, circumspice* (If you seek his monument, look around you). It is impossible to travel in Western Australia now without being touched by the work of C. Y. O'Connor. Few sections of our community—or indeed of Australia as a whole—have not benefited directly or indirectly from O'Connor's legacy: those who import or export goods from Fremantle; those who have ever arrived or departed from the port, or those who sail off shore from yacht clubs up river; those who live in the dry lands east of the Darling Range and depend on water from the pipeline for mining, agriculture and domestic use; those who travel about the State or across the Nullarbor by rail. All these have been O'Connor's beneficiaries. Since his death, all these modern facilities have been expanded and improved, but the foundation, the vision, and the original planning were the work of the Engineer-in-Chief. They are his rich legacy to us and to future generations.

C. Y. O'Connor was one of those great engineers of whom it could be said—borrowing Robert Drewe's words from his novel *The Drowner*, about an engineer on the goldfields in O'Connor's time—that he truly changed the order of things.

And that is as dramatic as life gets.

Portrait of O'Connor painted by his grand-daughter Frances from
a photograph, 1973.

Courtesy Judge V. J. O'Connor

Abbreviations used in Notes and Bibliography

BLP	Battye Library, Perth
NACC	National Archives, Christchurch
NZJHR	New Zealand Journal of the House of Representatives
PDWA	Parliamentary Debates, Western Australia
PROD	Public Records Office, Dublin
PWD	Public Works Department
RCBLD	Representative Church Body Library, Dublin
RLFC	Famine Relief Commission papers, Dublin
SHLF	Social History Library, Fremantle
SROWA	State Records Office, Western Australia
UWA	The University of Western Australia
V&P	Votes and Proceedings of the Western Australian Parliament
WCHMH	West Coast Historical Museum, Hokitika

Notes

INTRODUCTION

1. For the record: it is generally agreed that the Sydney Harbour Bridge was designed by the engineer J. J. Bradfield, although this is disputed by some authorities—see *Australian Dictionary of Biography* vol. VII, pp. 382–3; the head of the engineering team on the Snowy Mountains Hydro-electric Scheme was William Hudson, with the title Commissioner; the principal engineers responsible for the Forth Railway Bridge were John Fowler and Benjamin Baker, both knighted for their work by Queen Victoria; the Channel Tunnel involved ten big English and French construction companies within the trans-tunnel consortium and required so large a team of engineers at every stage that it is impractical to say who was responsible for the whole work.

2. Robert Drewe, *The Drowner*, Sydney, 1996, p. 206.

3. B. Trinder, *The Making of the Industrial Landscape*, London, 1982, p. 129.

4. A Memorial Committee selected Pietro Porcelli's design from seventeen entries Australia-wide and he received a commission of £1,500. Half the money was raised from a government donation; the other half by public subscription. The statue was first modelled in clay and later cast in bronze in Italy. Unveiled on 23 June 1911, it had two locations before finally being moved to its present site in 1974. See Reverend G. B. Keane, in *Early Days*, journal of the Royal Western Australian Historical Society, vol. 8, part 5, pp. 9–28.

5. Letter from Charlotte Brontë to her father, Patrick Brontë, London, 7 June 1851. Quoted in Juliet Barker, *The Brontës: A Life in Letters*, London, 1997.

6. Yvonne ffrench, *The Great Exhibition: 1851*, London, n.d., p. 215.

7. Quoted in G. Trevelyan's *Life & Letters of Macaulay*, vol. 2, London, 1876, p. 205.

8. G. M. Young, *Victorian England: Portrait of An Age*, London, 1986 [1936].

9. S. T. Coleridge, *On the Contributions of the Church and State*, London, 1972 [1830], p. 46.

10. Baron Meidinger, quoted in Patrick Howarth, *The Year Is 1851*, London, 1951, p. 40.

CHAPTER 1

1. William R. Wilde, *The Beauties and Antiquities of the Boyne*, facsimile edn Cork, 1978, p. 15. William Wilde was the father of poet and playwright Oscar Wilde.

[253]

2. Merab Tauman, in *The Chief: C. Y. O'Connor*, Nedlands, 1978, p. 5, states that local opinion suggested that Mrs Elizabeth O'Connor persuaded her husband to add the third storey to the original two-storey house. This seems unlikely to this writer after careful inspection, advice from a local architect working on the refurbishment, and the fact that an identical house by the same builder, the old Rectory, was three-storey. Evidence of the back wing being added at a later date is clearly visible.

3. It is unlikely that Martha Weld was related to the English Catholic Weld family, which produced a Governor of Western Australia.

4. Tauman, p. 4.

5. Curren, entry no. 363, 'Avonmore', quoted in George E. Cokayne, *Complete Peerage*, London, 1910, p. 302.

6. ibid.

7. Quoted in Cecil Woodham-Smith. *The Great Hunger: Ireland 1845–49*, London, 1964, p. 17.

8. O'Connor to Lomas, 1194A/10, BLP.

9. G. M. Young, *Victorian England: Portrait of An Age*, London, 1986 [1936], p. 14.

CHAPTER 2

1. Cecil Woodham-Smith, *The Great Hunger: Ireland 1845–49*, London, 1964, p. 38.

2. *Griffiths' Valuation of Tenements* [County of Meath], Dublin, 1854.

3. The Devon Commission, quoted in Woodham-Smith, p. 32.

4. Letting on conacre: the tenant farmer would let a small section of land to his worker for the season only, so that the poor worker could grow his own crop.

5. Quoted in Woodham-Smith, p. 30.

6. Each July and August, when the previous year's crop had been depleted and the new crop awaited, there was a period known as 'the meal months', when peasants, if they could afford it, had to purchase meal as an alternative.

7. R. D. Edwards & T. D. Williams (eds), *The Great Famine: Studies in Irish History 1845–1852*, Dublin, 1994, pp. 96–7.

8. RLFC, 2/Z14338, 25/10/1845, PROD.

9. RLFC, 2/Z470, 6/11/1845, PROD. Also see *Meath Herald*, 15 November 1845.

10. Charles S. Clements, *Alleged Dearth of Potatoes in Meath*, RLFC, 2/Z530, 8/1/1846, PROD.

11. ibid.

12. D. Cusack, *The Great Famine in Co. Meath*, Navan, 1996.

13. ibid.

14. Edwards & Williams, p. 217.

15. Quoted in *Meath Herald*, 28 March 1846.

16. ibid.
17. Edwards & Williams, p. 249.
18. Thomas O'Neill, quoted in ibid., p. 243.
19. Cusack.
20. Woodham-Smith.
21. Cusack, p. 5.
22. Merab Tauman, *The Chief: C. Y. O'Connor*, Nedlands, 1978, p. 5.
23. Edwards & Williams, p. xiii.
24. Tauman, p. 8.
25. Cited in Woodham-Smith, p. 399.

CHAPTER 3

1. J. O'Neill, 'Waterford's five railways', *Journal of the Irish Railway Record Society*, vol. 16, no. 101, Dublin, October 1986. See also R. V. Comerford, 'Ireland 1850–70: Post-famine and mid-Victorian', in W. E. Vaughan (ed.), *A New History of Ireland*, vol. V, *1801–1870*, Oxford, 1989, pp. 374–5.
2. ibid.
3. It is of passing interest to note that in Britain, where the nationalized railways were returned to private ownership under Mrs Thatcher's government in 1995, various companies run the trains on lines operated by the Railtrack Company, which makes a massive profit. The wheel has come full circle.
4. K. A. Murray & D. B. McNeill, *The Great Southern and Western Railway*, Dublin, 1976, p. 109.
5. Patrick Howarth, *The Year Is 1851*, London, 1951, p. 227.
6. William R. Wilde, *The Beauties and Antiquities of the Boyne*, facsimile edn Cork, 1978, p. 133.
7. Box Hill Tunnel is considered one of the most beautiful examples of Brunel's classical architecture. In *An Illustrated History of Civil Engineering*, London, 1964, p. 122, J. P. Pannell writes, 'surely not from coincidence, the rising sun shines through the tunnel every 9th April—Brunel's birthday'.
8. The biographical note on C. Y. O'Connor in *The Australian Encyclopedia*, vol. 6, Sydney, 1963, pp. 385–6, states that Charles's parents resisted his desire to study civil engineering, believing it to be too humble a profession for one of his class. There is no evidence extant for this view; rather, the opposite, as presented here.

CHAPTER 4

1. Michael Quane, *Bishop Foy's School, Waterford*, Cork Historical and Archaeological Society booklet, Cork, 1959, p. 104.

2. Register of Bishop Foy's School, 1711–1902, MS 523, RCBLD.
3. Medals for good conduct presented by the headmaster of Waterford Academy to George O'Connor, Charles's older brother, remain in the possession of Muriel Dawkins.
4. A documentary film on the life of C. Y. O'Connor screened at some time at the C. Y. O'Connor Museum, Mundaring, Western Australia, depicts the O'Connor house in Waterford as a poor, thatched cottage. There is no evidence that this was their home.
5. *A New System of Practical Domestic Economy*, London, 1824.
6. Quoted in Patrick Howarth, *The Year Is 1851*, London, 1951, pp. 82 and *passim*.
7. Michael Quane, *Waterford Corporation Free School*, Cork Historical and Archaeological Society booklet, Cork, 1959, pp. 82–103.
8. Several biographical notes on O'Connor state that he attended lectures at Trinity College but there is no record in the college archives of his having done so.
9. Quoted in B. Trinder, *The Making of the Industrial Landscape*, London, 1982, p. 129.
10. Obituary in *Minutes and Proceedings*, Institution of Civil Engineers, London, vol. 122, 1894–95, pp. 386–7.
11. Brunel, quoted in J. P. Pannell, *An Illustrated History of Civil Engineering*, London, 1964, p. 123.
12. O'Connor to Darley, 6 May 1865, O'Connor Papers, 3436A/38, BLP.
13. Frances O'Connor, letter, O'Connor Papers, 3436A/10, BLP.
14. The Lords interpreted the King George II Parliament Act 'that a marriage between a Catholic and a Protestant, if celebrated by a Roman Catholic priest, shall be deemed null and void'; in spite of this, the Lords referred to Yelverton, in one instance, as 'the scoundrel'.
15. This account of the Yelverton case is taken mainly from *The Times*, 5 March 1861, 31 July 1867; and *National Dictionary of Biography*, vol. XXI (Yelverton, William Charles) and vol. XII (Longworth, Maria Teresa), Oxford, 1953 [1903]. The *National Dictionary of Biography* notes that this affair is also fully reported in *The Yelverton Marriage Case*, and in Maria Longworth's *The Yelverton Correspondence, with Introduction and Connecting Narrative*, Edinburgh, 1863.
16. See Joan Weir, *Back Door to the Klondike: An Account of Viscount Algernon Yelverton's Adventures Gold Prospecting in 1898*, Ontario, 1988.

CHAPTER 5

1. Merab Tauman, *The Chief: C. Y. O'Connor*, Nedlands, 1978, p. 10.
2. Probate records, 1864, PROD.
3. *Griffiths' Valuation of Tenements* [County of Meath], Dublin, 1854.
4. See *The Australian Encyclopedia*, vol. 6, Sydney, 1963, pp. 385–6.

5.   O'Connor to Under Secretary, Public Works, 25 July 1898, PWD 7329/98, SROWA.

6.   Smith to Hemans, 3 December 1864, O'Connor Papers, BLP.

7.   *West Coast Times* (Hokitika), 22 March 1880, p. 2.

8.   O'Connor to Darley, 6 May 1865, O'Connor Papers, 3436A/38, BLP.

9.   William Malcolmson to C. Y. O'Connor, November 1864, O'Connor Papers, 3436A/3, BLP.

10. Smith to Hemans, O'Connor Papers, 3436A/2, BLP.

11. Letter from Viscount Avonmore, Roscrae, 8 December 1864, O'Connor Papers, 3436A/91, BLP.

12. O'Connor to Darley, op. cit.

CHAPTER 6

1.   O'Connor to Darley, 6 May 1865, O'Connor Papers, 3436A/38, BLP.

2.   ibid.

3.   ibid.

4.   W. H. Oliver & B. R. Williams (eds), *The Oxford History of New Zealand*, Oxford, 1981, p. 117.

5.   William Pember Reeves, *The Long White Cloud*, London, 1998 [1924], pp. 236–7.

6.   ibid., p. 39.

7.   O'Connor to Darley, op. cit.

8.   ibid.

9.   ibid.

10. ibid.

11. ibid.

12. Arthur Dudley Dobson, *Reminiscences*, London, 1930, p. 139.

13. Reeves, p. 229.

14. James Drummond, *The Life and Work of Richard John Seddon with a History of the Liberal Party in New Zealand*, London, 1907, p. 11.

15. Captain James Cook, *Journal 1768–71*, vol. 1, Canberra, 1999, p. 270.

16. 'Address of the Superintendent, Mr S. Bealey, upon the opening of the twenty-fourth session of the Prov. Council, 21 Nov. 1865', *J. Proc. Prov. Coun. Cant.*, 1865–66.

CHAPTER 7

1.   Arthur Dudley Dobson, *Reminiscences*, London, 1930, pp. 36–7.

2.   ibid., p. 65.

3.   William Pember Reeves, *The Long White Cloud*, London, 1998 [1924], p. 29.
4.   O'Connor diary, 1869, O'Connor Papers, 3436A/79, BLP.
5.   Reeves, p. 231.
6.   Dobson, pp. 121–3.
7.   Mueller, Gerhard, *My Dear Bannie: Letters from the West Coast 1865–6*, Christchurch, 1958, p. 177.
8.   Dobson, p. 159.
9.   *Kumara Times*, 14 May 1888; see also 28 December 1887.
10. ibid., 28 December 1887, p. 2.
11. Edward Dobson, 'Report of the Provincial Engineer of Public Works, 22 Oct. 1866', *J. Proc. Prov. Council*, 1865–66, p. 148.
12. Lord Lyttelton, quoted in William Downie Stewart, *William Rolleston: A New Zealand Statesman*, Wellington, 1940, p. 36.
13. Testimonial from E. Dobson, 23 August 1871, then Chief Engineer, Department of Victorian Water Supply, Melbourne, O'Connor Papers, 1994A/2, BLP.
14. Records in WCHMH show that O'Connor owned shares in the Hokitika–Greymouth Railway, as did most of the townspeople.
15. Diary, 1869.
16. Mueller.
17. Diary, 1869.
18. ibid.
19. *West Coast Times* (Hokitika), 22 March 1880.
20. The Hon. William Rolleston, tribute in *The Press*, 13 March 1902, O'Connor Papers, 3436A/43, BLP.

CHAPTER 8

1.   P. R. May, *Mines and Militants: Politics in Westland 1865–1918*, Christchurch, 1975, pp. 44–5.
2.   ibid.
3.   Illuminated address, County of Westland, original in possession of Judge V. J. O'Connor, Western Australia.
4.   May, p. 45.
5.   Geoffrey Rice, *Christchurch Changing: An Illustrated History*, Christchurch, 1999.
6.   Merab Tauman, *The Chief: C. Y. O'Connor*, Nedlands, 1978, p. 29.
7.   It was on this tour, while the Prince was visiting Sydney, that an attempt on his life was made by a Fenian.
8.   Kathleen O'Connor, 'Memoir of Her Father', typescript, n.d., O'Connor Papers, 3436A/90, BLP.
9.   Copies of these certificates are now in O'Connor Papers, BLP.
10. Kathleen O'Connor to Bridget O'Connor, n.d., in possession of Mrs Muriel Dawkins.

11. Sr Mary Raphael, quoted in P. Hutchings & J. Lewis, *Kathleen O'Connor: Artist in Exile*, Fremantle, 1987, p. 107.

12. Peter Porter, 'An Exequy', *Collected Poems*, London, 1988, pp. 246–7.

13. Kathleen O'Connor, 'Memoir of Her Father'.

14. *Lyttelton Times*, 25 August 1874.

15. William Pember Reeves, *The Long White Cloud*, London, 1998 [1924], p. 241.

16. Dillon Bell, quoted in Tauman, p. 31.

17. Outwards Correspondence, District Engineer, 1877–88, Acc. 125, NACC.

18. ibid.

19. ibid.

20. ibid.

CHAPTER 9

1. O'Connor Papers, 4690A/4, BLP.

2. Merab Tauman, *The Chief: C. Y. O'Connor*, Nedlands, 1978, p. 32.

3. Kathleen O'Connor, 'Memoir of Her Father', typescript, n.d., O'Connor Papers, 3436A/90, BLP.

4. William Rolleston, *The Press*, 13 March 1902, O'Connor Papers, 3436A/43, BLP.

5. Quoted in Tauman, p. 32.

6. O'Connor Papers, 4690A/3, BLP. See also 'Outwards Correspondence, District Engineer, 1877–88, Acc. 125, NACC.

7. Tauman, p. 33.

8. 4690A/2, BLP. See also Outwards Correspondence.

9. O'Connor Papers, 4690A/1, BLP.

10. Outwards Correspondence.

11. Kate O'Connor, 'Reminiscences'.

12. See Jeanine Graham, 'The pioneers', in Keith Sinclair (ed.), *Oxford Illustrated History of New Zealand*, Auckland, 1997, p. 72.

13. *West Coast Times*, 23 March 1880.

14. Inquest Book, 1865–80, Acc. 306, NACC.

15. *West Coast Times*, 22 March 1880.

16. ibid., 7 January 1880.

17. ibid., 16 January 1880.

18. O'Connor Papers, 1994A/9, BLP.

19. See *Western Mail*, 5 August 1898, p. 41, attacks by Bargigli and O'Connor's reply. Other later attacks in the *Sunday Times* are noted in Part III.

20. Hunter McAndrew quoted in Tauman, p. 33.

21. 'The Inspecting Engineer, Middle Island, to the Commissioners, 30 December 1882', report in NZJHR, vol. 11, D.2, pp. 118, 69–75.

22. Rolleston to O'Connor, 19 November 1883, O'Connor Papers, 1994A/3, BLP.

CHAPTER 10

1. Letter, n.d., O'Connor Papers, 1994A/2, BLP.
2. For this quote, and other information about Mrs Swainson's school, I am indebted to author Julie Lewis, in P. Hutchings & J. Lewis, *Kathleen O'Connor: Artist in Exile*, Fremantle, 1987, p. 15.
3. O'Connor Papers, 1994A/3, BLP.
4. Julius Vogel, letter dated 29 December 1884, O'Connor Papers, 1994A/3, BLP.
5. 14 December 1884, O'Connor Papers, 3436A/7, BLP.
6. 3 November 1884, O'Connor Papers, 3436A/8, BLP.
7. Personal memorandum in O'Connor's handwriting, unaddressed, dated 30 April 1991, O'Connor Papers, 1994A/7, BLP.
8. ibid.
9. O'Connor's draft report to parliament, 20 August 1885, O'Connor Papers, 1994A/7, BLP.
10. ibid.
11. Rolleston to O'Connor, 19 November 1883, O'Connor Papers, 1994A/3, BLP.
12. Blackett to O'Connor, 23 February 1884, O'Connor Papers, 1994A/2, BLP.
13. O'Connor Papers, 1994A/2, BLP.
14. *Evening Star* (Wellington), 17 March 1889.
15. T. Fergus, Minister for Public Works, to O'Connor, 10 May 1890, O'Connor Papers, 1994A/8, BLP.
16. ibid.
17. Merab Tauman, *The Chief: C. Y. O'Connor*, Nedlands, 1978, p. 40.
18. O'Connor to T. Fergus, 12 May 1890, O'Connor Papers, 1994A/8, BLP.
19. T. Fergus to O'Connor, op. cit.

CHAPTER 11

1. 'Gisborne Harbour: Report by the Marine Engineer Together with Copies of Drawings', NZJHR, 1891, D.3.
2. Letter of commendation from Timaru Harbour Board, 3 April 1891, included among O'Connor's testimonials submitted to Forrest, April 1891, O'Connor Papers, BLP.
3. O'Connor Papers, 3436A/7, 3436A/8, BLP.
4. F. K. Crowley, 'Forrest the Politician 1891–1918', unpublished manuscript, Battye Library, p. 12.
5. O'Connor Papers, 1994A/7, BLP.
6. ibid.

7. James McKerrow to John Forrest, 23 April 1891, O'Connor Papers, 1194A/10, BLP.

8. F. K. Crowley, *Big John Forrest*, Nedlands, 2000, p. 87.

9. The three striking words on a cable from Forrest to O'Connor form the title of one of the chapters of Merab Tauman's biography (*The Chief: C. Y. O'Connor*, Nedlands, 1978) and are quoted in several short accounts of O'Connor's life and work. The cable is preserved in the O'Connor Papers, 1994A/7, BLP.

10. O'Connor Papers, 1994A/7, BLP.

11. ibid.

12. ibid.

13. ibid.

14. ibid.

15. ibid.

16. ibid.

CHAPTER 12

1. Merab Tauman (*The Chief: C. Y. O'Connor*, Nedlands, 1978) and others state that only Aileen arrived in Western Australia with her father on the *Massilia*. Passenger lists for Albany arrivals confirm that the family was divided in this way; SROWA.

2. *Age* (Melbourne), shipping report, 21 May 1891.

3. C. Y. O'Connor to John Lomas, Christchurch, 6 January 1894, O'Connor Papers, BLP.

4. Kathleen O'Connor, 'Memoir of Her Father', typescript, n.d., O'Connor Papers, 3436A/90, BLP.

5. ibid.

6. 'Cinderella State' first appears in a cartoon in the *Bulletin*, 4 August 1888.

7. John Boyle O'Reilly, *Songs of the Southern Seas*, p. 137, quoted in A. G. Evans, *Fanatic Heart*, Nedlands, 1997, p. 133.

8. Anthony Trollope, *South Australia, Western Australia and N.Z.*, London, 1884.

9. *West Australian*, 2 June 1891.

10. Albert Calvert, *My Fourth Tour of Western Australia*, Perth, 1989, p. 10.

11. See F. K. Crowley, *Big John Forrest*, Nedlands, 2000, n. 111 concerning Forrest's title, p. 522.

12. R. E. N. Twopenny, *Town Life in Australia*, London, 1883, p. 169.

13. *West Australian*, 13 May 1891.

14. Ray & John Oldham, *George Temple-Poole: Architect of the Golden Years 1885–1897*, Nedlands, 1980.

15. *West Australian*, 3 June 1891.

CHAPTER 13

1. For a detailed description of the new government offices at that time, see *Year Book of Australia 1893*, pp. 837–8. The Public Works Department later moved to the vacated Pensioner Barracks at the west end of St Georges Terrace, but not until after the death of C. Y. O'Connor.

2. For a note on civil servants and other workers travelling to work in the late nineteenth century, see C. T. Stannage, *The People of Perth*, Perth, 1979, pp. 133–4.

3. Perth's railway station in the 1890s was a little to the west of its present position.

4. A small section of the Long Jetty has been preserved south of Arthur Head at what is known today as Bathers Beach.

5. Quoted in J. K. Ewers, *The Western Gateway*, 2nd edn, Nedlands, 1971 [1948], p. 244.

6. Report of the Harbour and Light Department, 1889–90, SROWA.

7. *West Australian*, 2 June 1891.

8. Sir John Coode, 'Report on the Question of Harbour Works at Fremantle, 18 March 1887', V&P, 1887, Parliamentary Paper 18.

9. ibid.

10. Accidents were frequently reported on the Eastern line, crossing the Darling Range. One of the most spectacular occurred in December 1885 when the brakes of a train descending the Greenmount Hill towards Perth could not hold the heavy ballast wagons behind it. The *West Australian* reported that 'the train appears to have kept on its mad course increasing in velocity every minute until it was shooting along a rate of one hundred and twenty miles an hour'. The engine driver and fireman leapt for their lives moments before the train broke from the rails with such force that it shot into the air and ploughed into the embankment. O'Connor saw immediately that these accidents were due to bad track design, light engines and too steep gradients.

11. Kathleen O'Connor, 'Memoir of Her Father', typescript, n.d., O'Connor Papers, 3436A/90, BLP.

12. Fremantle Rate Books, 1891–1900, SHLF.

13. Maureen F. Coghlan, 'Monuments in Stone', thesis, Claremont Teachers College, 1958, BLP. Sadly, 'Park Bungalow', which could have become a Fremantle museum to C. Y. O'Connor, was demolished in the early 1960s. A good impression of the style and atmosphere of the interior of 'Park Bungalow' can be gained from a visit to 'Samson House' in Ellen Street, diagonally opposite across the park. Owned by the National Trust and open to the public on Sundays, it was built around the same period. The O'Connors were frequent guests of the then owners, Michael and Mary Samson.

14. Kathleen O'Connor, op. cit.

15. ibid.

16. ibid.
17. Lucius C. Manning to Merab Tauman, 2 October 1972, in Tauman, *The Chief: C. Y. O'Connor*, Nedlands, 1978.
18. *West Australian*, 8 July 1891.
19. Kathleen O'Connor, op. cit.

CHAPTER 14

1. Governor's Speech at the opening of the First Parliament in January 1891, quoted in T. Manford, 'A History of Rail Transport Policy in Western Australia 1870–1911', PhD thesis, UWA, 1976, pp. 171–2 and *passim*.
2. F. K. Crowley, *Big John Forrest*, Nedlands, 2000.
3. PDWA, vol. IV, 1893, p. 126.
4. ibid., p. 127.
5. G. Spenser Compton, 'Yilgarn and Coolgardie railways', in *Early Days*, journal of the Western Australian Historical Society, vol. 5, part 2, p. 31.
6. ibid.
7. ibid.
8. PDWA, vol. IV, 1893, p. 117.
9. ibid.
10. PDWA, vol. VII, 1894, p. 1493.
11. *West Australian*, 22 February 1892.
12. PDWA, vol. III, 1893, p. 562.
13. V&P, 1896, vol. II; *Morning Herald*, 8 January 1896.
14. 'Report on the Working of Government Railways', 1891, V&P, 1893.
15. 'Report on the Working of Government Railways, 1892–5 by the Engineer-in-Chief', 1895, V&P. In his summary, he states: 'During the year 1891 the gross profit on the railways (exclusive of the Cossack–Roebourne Tramway) after deducting working expenses, was only £498, equivalent to only 0.05 per cent on the capital cost, leaving practically the whole of the interest on cost to be borne by the General Revenue; whereas during the year ended 30th June last, the earnings, in excess of working expenses, amounted to £113,954, which is equivalent to 5.44 per cent on the capital cost'.
16. *West Australian*, 17 March 1892.

CHAPTER 15

1. *West Australian*, 29 October 1891.
2. 'Correspondence Respecting a Safe and Commodious Harbour at Fremantle', no. 17, V&P, 1891.

3.   'Fremantle Harbour Works: Report by the Engineer-in-Chief upon the Entrance to the Estuary at the Mouth of the Swan River or at Rocky Bay', V&P, 1892, p. 4.

4.   Forrest to Venn, 30 September 1891, PWD 11962/19, SROWA.

5.   Venn to Forrest, 11 September 1891, PWD 1886/91, SROWA.

6.   'Fremantle Harbour Works', op. cit., p. 5.

7.   PDWA, 1892, vol. II, p. 326.

8.   ibid., pp. 315–27.

9.   ibid.

10.   Truthful Thomas, *Through the Spyglass*, Perth, 1905.

11.   This and all the previous verbatim extracts are taken from the 'Report of the Joint Select Committee on the Question of Harbour Works at Fremantle, 1891', V&P, 1891–92.

12.   *West Australian*, 17 November 1892.

13.   PDWA, 1891–92, op. cit.

14.   Merab Tauman, *The Chief: C. Y. O'Connor*, Nedlands, 1978, p. 239.

15.   O'Connor to Venn, 4 March 1892, WAS 1375, SROWA.

16.   ibid.

17.   Venn to O'Connor, 7 March 1892, WAS 1375, SROWA.

18.   See Frank Stevens, 'Rise of a great port', supplement to *Western Mail*, 6 June 1929, pp. 6–7.

19.   Tauman, p. 247.

20.   Kathleen O'Connor, 'Memoir of Her Father', typescript, n.d., O'Connor Papers, 3436A/90, BLP.

CHAPTER 16

1.   Albert Gaston, *Coolgardie Gold*, facsimile edn Perth, 1984, pp. 18–19.

2.   Between the years 1891 and 1900, there were 3,567 cases of typhoid and 301 deaths in Coolgardie; 1,570 cases and 189 deaths in Kalgoorlie; and in Western Australia overall, 15,854 cases and 1,642 deaths. See Vera Wittington, *Gold and Typhoid: Two Fevers*, Nedlands, 1988.

3.   *West Australian*, 26 September 1892.

4.   ibid., 24 September 1892.

5.   ibid.

6.   O'Connor Papers, 1194A/11, BLP.

7.   PDWA, 19 December 1892.

8.   *West Australian*, 17 November 1892.

9.   Kathleen O'Connor, 'Memoir of Her Father', typescript, n.d., O'Connor Papers, 3436A/90, BLP.

10. PDWA, 1893, vol. IV, pp. 117–27.
11. C. Y. O'Connor, 'Fremantle Harbour', in *Catalogue of Exhibits in the West Australian Section of the Paris Universal Exhibition of 1900*, Perth, 1900.
12. ibid.
13. *West Australian*, 31 May 1893.

CHAPTER 17

1.   *Inquirer*, 2 February 1894.
2.   ibid., 16 February 1894.
3.   *West Australian*, 27 January 1902.
4.   *Inquirer*, 11 May 1894.
5.   ibid.
6.   M963/93, SROWA.
7.   For a detailed review of water shortage on the Eastern line, see J. S. H. Le Page, *Building a State: The Story of the Public Works Department of Western Australia 1929–1985*, Perth, 1986, pp. 264–72.
8.   Merab Tauman, *The Chief: C. Y. O'Connor*, Nedlands, 1978, p. 121.
9.   *Inquirer*, 28 September 1894.
10. Debate on proposed removal of the railway workshops, 13 November 1894, PDWA, 1894, vol. VII.
11. *West Australian*, 16 March 1895.
12. PDWA, 1894, vol. VII.
13. ibid.
14. ibid.
15. *West Australian*, 9 February 1895.
16. *Inquirer*, 27 July 1894.
17. ibid.
18. *West Australian*, 21 March 1895.
19. ibid., 20 August 1895.
20. ibid.
21. ibid.
22. ibid.
23. ibid., 21 August 1895.
24. ibid.
25. Kathleen O'Connor, 'Memoir of Her Father', typescript, n.d., O'Connor Papers, 3436A/90, BLP.
26. Frances O'Connor to C. Y. O'Connor, 5 August 1894, O'Connor Papers, BLP.
27. *Western Mail*, 4 May 1895.

CHAPTER 18

1.   *West Australian*, 8 March 1894.
2.   F. Alexander, F. K. Crowley & J. D. Legge, *Origins of the Eastern Goldfields Water Scheme in Western Australia*, Nedlands, 1953, pp. 20–1.
3.   ibid.
4.   ibid.
5.   PDWA, 1895, vol. VIII, p. 1296.
6.   ibid., p. 1295.
7.   *West Australian*, 10 September 1895.
8.   ibid.
9.   Alexander et al., pp. 40–1. See also Merab Tauman, *The Chief: C. Y. O'Connor*, Nedlands, 1978, p. 149.
10. PWD 8575/96, SROWA.
11. It was government policy to send and receive cables in code in order to keep government business confidential, and possibly also to compress the content, and hence the cost, of messages.
12. C. Y. O'Connor, *Report on Proposed Water Supply (by Pumping) from Reservoirs in the Greenmount Ranges*, Perth, 1896.
13. *West Australian*, 13 September 1895.
14. 'Report on the Working of Government Railways 1892–1895', V&P, 1895, BLP.
15. *West Australian*, 2 September 1895.
16. 'Report on the Working of Government Railways'.
17. *West Australian*, 12 September 1895.
18. ibid., 26 April 1895.
19. ibid., 24 August 1895.
20. O'Connor had arranged for the *Premier* to be refitted in Adelaide and towed to Fremantle while on his visit to Albany. See *West Australian*, 19 October 1895.
21. *West Australian*, 15 November 1895.
22. PDWA, 1895, vol. VIII, p. 144.
23. *West Australian*, 2 March 1896.
24. *West Australian* and *Morning Herald*, correspondence 10, 11, 13, 17 March 1896.
25. ibid.
26. Quoted in F. K. Crowley, 'Forrest the Politician 1891–1919', manuscript, BLP.
27. PDWA, 1890–91, vol. I, p. 191.
28. Quoted in G. C. Bolton, *Alexander Forrest: His Life and Times*, Melbourne, 1958, p. 143.
29. Carruthers to O'Connor, 24 April 1896, 619/vol. 1, 2575/96, SROWA.
30. Hodgson to O'Connor, ibid.
31. PDWA, 1896, vol. IX, pp. 1–3.
32. ibid., pp. 130–51.
33. O'Connor to Carruthers, 3 March 1896, 619/vol. 1, SROWA.

34. Forrest to Piesse, 9 November 1896, ibid.
35. ibid.
36. O'Connor to Piesse, 14 November 1896, ibid.
37. O'Connor to Carruthers, 31 December 1896, ibid.
38. Alexander et al.

CHAPTER 19

1.   Geoffrey Blainey, 'The torment of the Water King', in *The Golden Mile*, Sydney, 1993.
2.   C. Y. O'Connor, *Report on Proposed Water Supply (by Pumping) from Reservoirs in the Greenmount Ranges*, Perth, 1896, p. 3.
3.   PDWA, 1896, vol. IX, pp. 267–72.
4.   ibid., p. 272.
5.   *Kalgoorlie Miner*, 3 August 1896.
6.   Blainey, p. 69.
7.   ibid.
8.   ibid.
9.   ibid., p. 70.
10. ibid.
11. ibid., p. 71.
12. O'Connor, p. 4.
13. Pers. comm., R. Hillman, K. Kelsall & H. Hunt, engineers for many years with PWD (later Water Authority of Western Australia, now Water Corporation), and now retired, July 2000.
14. Memorandum, 31 July 1894, O'Connor to F. Reed, 1478, SROWA.
15. R. H. B. Kearns, *Broken Hill: A Pictorial History*, Hawthorndene, South Australia, 1892, p. 209.
16. ibid., p. 214.
17. ibid.
18. PDWA, 1896, vol. IX, pp. 130–51.

CHAPTER 20

1.   *West Australian*, 17 September 1897, p. 5.
2.   ibid.
3.   ibid.
4.   G. B. Shaw, *Our Theatres in the Nineties*, vol. 3, London, 1954, p. 179.
5.   O'Connor Papers, BLP.
6.   *West Australian*, 5 May 1897.

7.   ibid.

8.   *West Australian*, 17 September 1897.

CHAPTER 21

1.   Hodgson to O'Connor, Acc. 689, AN7, vol. 1, 1898, SROWA.

2.   *West Australian*, 4 August 1897.

3.   PDWA, 1897, vol. XI, p. 246.

4.   PDWA, 1899, vol. XV, p. 1898.

5.   Quoted in Manford, op. cit., pp. 229–30.

6.   For a full report of the inquiry into the charges against John Davies, see V&P, 1901, vol. III, p. 358, Parliamentary Paper 40. See also *West Australian*, 30 October 1901.

7.   Forrest to Piesse, Acc. 689, AN7, vol. 1, 1898, SROWA.

8.   Letter signed 'Johnnie', *West Australian*, 17 June 1898.

9.   *Morning Herald*, 9, 16 June 1898; *West Australian*, 17 June 1898.

10.   C. Y. O'Connor to Piesse, 27 April 1898, Acc. 689, AN7, vol. 1, 1898, SROWA.

11.   O'Connor to Hodgson, Acc. 689, AN7, vol. 1, 1898, SROWA.

12.   18 February 1898, Acc. 689, AN7, vol. 1, 1898, SROWA.

13.   O'Connor to Hodgson, Acc. 689, AN7, vol. 1, 1898, SROWA.

14.   PDWA, 1898, vol. XIII, p. 2174.

15.   James Mephan Ferguson, *Mephan Ferguson: A Biography*, Broken Hill, 1992, p. 19.

16.   Minutes of the Acclimatization Committee, 17 July 1896 to 21 February 1902, Acc. 161, 254/1, BLP.

17.   It may seem surprising that O'Connor praised Marmion so warmly when it is remembered that Marmion was a critic of the Fremantle Harbour scheme and a persistent questioner during the harbour inquiry. He also led the delegation in protest at the removal of the Fremantle workshops. *West Australian*, 24 February 1898.

18.   Gertrude & Anthony Elworthy, *A Power in the Land: Churchill Julius 1847–1938*, Christchurch, 1971, p. 127.

19.   ibid.

20.   O'Connor to Lomas, 6 January 1894, O'Connor Papers, 1194A10, BLP.

21.   ibid.

22.   Truthful Thomas, *Through the Spyglass*, Perth, 1905.

23.   ibid.

24.   PDWA, 1898, vol. III, p. 1728.

25.   *Sunday Times*, 2 December 1900.

26.   *West Australian*, 22 May 1899.

27.   *Sunday Times*, 22 October 1899.

CHAPTER 22

1. n.d., O'Connor Papers, 3436A/13, BLP.
2. Charles Stuart Russell Palmer, *Coolgardie Water Supply*, with extract of discussion and minutes of the Institution of Civil Engineers, London, 1905, p. 18.
3. *Sunday Times*, 2 December 1900.
4. ibid., 23 December 1900.
5. ibid., 21 August 1900.
6. ibid.
7. ibid., 19 August 1900.
8. T. Manford, 'A History of Rail Transport in Western Australia 1870–1911', MA thesis, UWA, 1976, p. 279.
9. PDWA, 1900, vol. XVII, pp. 55–6.
10. PDWA, 1900, vol. XVIII, p. 1387.
11. *West Australian*, 2 January 1901.
12. ibid., 24 January 1901.
13. ibid., 8 February 1901.
14. ibid., 8 April 1901.
15. ibid., 9 April 1901.
16. ibid., 11 April 1901.
17. ibid., 19 April 1901.
18. ibid., 10 August 1901.
19. 'Report of the Royal Commission upon the Conduct and Completion of the Coolgardie Water Scheme', p. 300, SROWA.
20. Couston to Hodgson, papers in connection with Coolgardie Water Supply, V&P, 1901–02, vol. IV, A28, pp. 3–4.

CHAPTER 23

1. Couston to Hodgson, papers in connection with Coolgardie Water Supply, V&P, 1901–02, vol. IV, A28, pp. 3–4.
2. Truthful Thomas, *Through the Spyglass*, Perth, 1905.
3. V&P, 1901–02, vol. IV, A28, pp. 3–4.
4. PDWA, 1902, vol. XX, p. 3155.
5. Truthful Thomas.
6. PDWA, 1902, vol. XX, p. 2704.
7. ibid., p. 2701.
8. Merab Tauman, *The Chief: C. Y. O'Connor*, Nedlands, 1978, p. 220.
9. V&P, 1901–02, vol. IV, A28, pp. 3–4.
10. Tom Hungerford, reminiscences to the author, 13 July 2000.
11. Coroner's Report, 11 March 1902, Acc. 997, SROWA.

12. Lady Lee-Steere, quoted in Tauman, p. 226.
13. V&P, 1901–02, vol. IV, A28, pp. 3–4.
14. PDWA, 1902, vol. XX, p. 3155.
15. Sanderson, quoted in Tauman, p. 227.
16. ibid.
17. Charles Stuart Russell Palmer, *Coolgardie Water Supply*, with extract of discussion and minutes of the Institution of Civil Engineers, London, 1905, p. 161.
18. Coroner's Report, op. cit.
19. Jull held the same post in Western Australia that O'Connor had held in Wellington, possibly a factor in their friendship. Like O'Connor, Jull was a man of integrity and quiet efficiency. He became the first head of the public service in Western Australia.
20. Coroner's Report, op. cit.
21. ibid.
22. ibid.

CHAPTER 24

1. Treasury Papers, 1902, Acc. 1496, SROWA.
2. ibid.
3. ibid.
4. ibid.
5. Coroner's Report, 11 March 1902, Acc. 997, SROWA.
6. ibid.
7. *West Australian*, 12 March 1902.
8. O'Connor Papers, 3436A/43, BLP.
9. Cutting, n.d., in ibid.
10. *The Engineer*, 18 April 1902, cutting in ibid.
11. *Sunday Times*, 16 March 1902.
12. Schedule of Assets and Liabilities of C. Y. O'Connor, quoted in Merab Tauman, *The Chief: C. Y. O'Connor*, Nedlands, 1978, pp. 252–3.
13. PDWA, 1902, vol. XXI, pp. 1127–30.
14. ibid.
15. Roderick O'Connor's Account Book, in possession of V. J. O'Connor, Perth.
16. Kathleen O'Connor, 'Memoir of her Father', typescript, n.d., O'Connor Papers, 3436A/90, BLP.
17. Transcript in the O'Connor Papers, BLP.
18. O'Connor Papers, BLP.
19. V&P, 1902, pp. 1566–840.
20. ibid.
21. H. E. Hunt, 'Chronological Examination of Design and Construction of the

Coolgardie Pipeline', and other unpublished papers prepared for the National Trust of Western Australia, 2000. See also Charles Stuart Russell Palmer, *Coolgardie Water Supply*, with extract of discussion and minutes of the Institution of Civil Engineers, London, 1905, pp. 44–8.

22. Hunt.

23. 'Report of the Royal Commission upon the Conduct and Completion of the Coolgardie Water Scheme', SROWA.

24. J. S. Battye, *The Cyclopaedia of Western Australia*, Adelaide, 1912–13.

25. Reported in *West Australian*, 26 January 1903.

26. Palmer, p. 28.

27. *Morning Herald*, 26 January 1903.

28. *West Australian*, 26 January 1903.

29. ibid.

30. *Morning Herald*, 26 January 1903.

31. *West Australian*, 26 January 1903.

32. W. H. Auden, 'The Musée de Beaux Arts', in *Collected Shorter Poems 1927–1957*, London, 1969, p. 123.

EPILOGUE

1. *West Australian*, 24 November 1999.

2. Martyn Webb, 'The death of a hero: The strange suicide of Charles Yelverton O'Connor', *Early Days*, Royal Western Australian Historical Society journal, vol. II, part 1, 1985, pp. 104–5.

3. Merab Tauman, *The Chief: C. Y. O'Connor*, Nedlands, 1978, p. 251.

4. *West Australian*, 13 May 1901, and others.

5. 'Report of the Royal Commission upon the Conduct and Completion of the Coolgardie Water Scheme', vol. 2, pp. 257–60, 362–9, SROWA.

6. Yang & Clum, quoted in Elizabeth Greenwood, 'Suicide in Adults', research paper, 1966.

7. *The Engineer*, 18 April 1902 (cutting in O'Connor Papers, BLP).

Bibliography

PRIMARY SOURCES

Archival Sources

C. Y. O'Connor Papers in the John Forrest Collection, MN477, BLP.

Coroner's Report, Crown Law Department File No. 976/1902, Acc. 997, SROWA.

Kathleen O'Connor Archive, BLP.

Kathleen O'Connor's Red Scrapbook, O'Connor Papers, BLP.

Kathleen O'Connor, 'Memoir of Her Father', typescript, n.d., O'Connor Papers, 3436A/90, BLP.

O'Connor Family Papers (in part on microfilm), MN403, Acc. 1994A–3436A, Private Archive Stack, BLP.

Public Works Department Files, 1891–1902, SROWA.

Public Works Department Papers, NACC.

Votes and Proceedings of the Western Australian Parliament, 1891–1902, BLP.

West Australian Year Books and Post Office Books, 1891–1902, BLP.

Newspapers

Age (Melbourne), 1891.

Australian Advertiser (Albany), 1891–1902.

Daily News (Perth), 1891–1902.

Evening Post (Wellington), 1881–91.

Evening Star (Wellington), 1889.

Inquirer (Perth), 1891–1902.

Kalgoorlie Miner, 1896–1902.

Kumara Times, 1871–80.

Lyttelton Times, 1874.

Meath Herald, 1845–50.

Morning Herald (Perth), 1896–1902.

Scotsman (Edinburgh), 1862–66.

Sunday Times (Perth), 1891–1902.

The Times (London), 1834–68.

Waterford Chronicle, 1851–60.

Waterford News, 1851–60.

West Australian (Perth), 1891–1902.

West Coast Times (Hokitika), 1871–80.
Western Mail (Perth), 1891–1902.

Family Members Consulted

Dawkins, Mr John
Dawkins, Mrs M.
Dunlop, Mr Stewart
Lee-Steere, Sir Ernest
Nuttall, Mrs P.
O'Connor, Judge V. J.

SECONDARY SOURCES

Published Works

Adam, James, *Twenty-five Years of Emigrant Life in the South of New Zealand*, Wellington, n.d.
Alexander, F., F. K. Crowley & J. D. Legge, *Origins of the Eastern Goldfields Water Scheme in Western Australia*, Nedlands, 1953.
Australian Encyclopaedia, The, vol. 6, Sydney, 1963.
Australian Dictionary of Biography, vol. 7, *1891–1939*, Melbourne, 1979, pp. 51–4.
Ayris, Cyril, *C. Y. O'Connor: The Man for the Time*, Perth, 1996.
Barker, Juliet, *The Brontës: A Life in Letters*, London, 1997.
Battye, J. S., *The Cyclopaedia of Western Australia*, Adelaide, 1912–13.
Blainey, G., *The Golden Mile*, Sydney, 1993.
Bolton, G. C., *Alexander Forrest: His Life and Times*, Melbourne, 1958.
Calvert, Albert, *My Fourth Tour of Western Australia*, Perth, 1989.
Cambridge History of the British Empire, vol. VII, part 2, *New Zealand*, London, 1933.
Cokayne, George E., *Complete Peerage*, London, 1910.
Colebatch, Sir Hal, *A Story of a Hundred Years, 1829–1929*, Perth, 1929.
Coleridge, S. T., *On the Contribution of the Church and State*, London, 1972 [1830].
Compton, G. Spenser, 'Yilgarn and Coolgardie railways', *Early Days*, journal of the Royal Western Australian Historical Society, vol. 5, part 2.
Cook, Captain James, *Journal 1768–71*, Canberra, 1999.
Crowley, F. K., *Australia's Western Third*, Melbourne, 1960.
—— *Big John Forrest 1847–1918*, Nedlands, 2000.
—— *Forrest 1847–1918*, vol. 1, *1847–91: Apprenticeship to Premiership*, St Lucia, 1971.
Cusack, D., *The Great Famine in Co. Meath*, Navan, 1996.
Cyclopedia of New Zealand, Christchurch, 1906.
Dobson, Arthur Dudley, *Reminiscences*, London, 1930.
Drewe, Robert, *The Drowner*, Sydney, 1996.

Drummond, James, *The Life and Work of Richard John Seddon with a History of the Liberal Party in New Zealand*, London, 1907.

Edwards, R. D. & T. D. Williams (eds), *The Great Famine: Studies in Irish History 1845–1852*, Dublin, 1994.

Elworthy, Gertrude & Anthony, *A Power in the Land: Churchill Julius 1847–1938*, Christchurch, 1976.

Evans, A. G., *Fanatic Heart: A Life of John Boyle O'Reilly 1844–1890*, Nedlands, 1997.

Ewers, J. K., *The Story of the Pipe-line*, Perth, 1935.

—— *The Western Gateway*, 2nd edn, Nedlands, 1971 [1948].

Ferguson, James M., *Mephan Ferguson: A Biography*, Broken Hill, 1992.

ffrench, Yvonne, *The Great Exhibition: 1851*, London, n.d.

Finch, J. K., *Engineering and Western Civilisation*, London, 1951.

Furkett, F. W., *Early New Zealand Engineers*, Wellington, 1953.

Gaston, Albert, *Coolgardie Gold*, London, 1937, facsimile edn Perth, 1984.

Graham, Jeanine, 'The pioneers', in Keith Sinclair (ed.), *Oxford Illustrated History of New Zealand*, Auckland, 1997.

Griffiths' Valuation of Tenements, Dublin, 1854.

Harris, M., 'Charles Yelverton O'Connor, engineer-economist', *University Studies in History and Economics*, vol. 1, no. 1, UWA, 1934.

Hartley, Richard, *A Guide to Printed Sources for the History of the Eastern Goldfields Region of Western Australia*, Nedlands, 2000.

Hasluck, Alexandra, *C. Y. O'Connor*, Great Australians Series, Melbourne, 1965.

Hetherington, E., *Handbook for Intending Emigrants*, London, 1882.

Hitchcock, J. K., *A History of Fremantle 1829–1929*, Fremantle, 1929.

Howarth, Patrick, *The Year Is 1851*, London, 1951.

Hutchings, P. & J. Lewis, *Kathleen O'Connor: Artist in Exile*, Fremantle, 1987.

Illustrated Handbook of Western Australia, Royal Commission for the Paris Exhibition, Perth, 1900.

Keane, Reverend G. B., in *Early Days*, journal of the Royal Western Australian Historical Society, vol. 8, part 5, pp. 9–28.

Kearns, R. H. B., *Broken Hill: A Pictorial History*, Hawthorndene, South Australia, 1892.

Kinealy, Christine, *This Great Calamity: The Irish Famine 1845–52*, Dublin, 1995.

Kirby, R. S., *Early Years of Modern Civil Engineering*, London, 1932.

Le Page, J. S. H., *Building a State: The Story of the Public Works Department of Western Australia 1829–1985*, Perth, 1986.

Longworth, Maria, *The Yelverton Correspondence, with Introduction and Connecting Narrative*, Edinburgh, 1963.

Lyons, F. J., *Ireland Since the Famine*, London, 1971.

May, Philip Ross, *Hokitika, Goldfields Capital*, Christchurch, 1964.

—— *Miners and Militants: Politics in Westland 1865–1918*, Christchurch, 1975.

Millar, J. Hackett, *Westland's Golden Sixties*, Wellington, 1959.

Mueller, Gerhard, *My Dear Bannie: Letters from the West Coast 1865–6*, Christchurch, 1958.

Murray, K. A. & D. B. McNeill, *The Great Southern and Western Railway*, Dublin, 1976.

National Dictionary of Biography to 1900, Oxford, 1953 [1903].

New System of Practical Domestic Economy, A, London, 1824.

O'Connor, C. Y., *Report on Proposed Water Supply (by Pumping) from Reservoirs in the Greenmount Ranges*, Perth, 1896.

—— 'Fremantle Harbour', in *Catalogue of Exhibits in the Western Australian Section of the Paris Universal Exhibition of 1900*, Perth, 1900.

O'Gráda, Cormac, *Black 47 and Beyond: The Great Irish Famine in History, Economy and Memory*, Princetown, 1999.

Oldham, Ray & John, *George Temple-Poole: Architect of the Golden Years, 1885–1897*, Nedlands, 1980.

Oliver, W. H., & Williams, B. R. (eds), *The Oxford History of New Zealand*, Oxford, 1981.

O'Neill, J., 'Waterford's five railways', *Journal of the Irish Railway Record Society*, vol. 16, no. 101, Dublin, October 1986.

Palmer, Charles Stuart Russell, *Coolgardie Water Supply*, with extract of discussion and minutes of the Institution of Civil Engineers, London, 1905.

Pannell, J. P., *An Illustrated History of Civil Engineering*, London, 1964.

Pierre, W. A., *Canterbury Provincial Railways*, Christchurch, 1962.

Quane, Michael, *Bishop Foy's School, Waterford*, Cork Historical and Archaeological Society booklet, Cork, 1959.

—— *Waterford Corporation Free School*, Cork Historical and Archaeological Society booklet, Cork, 1959.

Reeves, William Pember, *The Long White Cloud*, London, 1998 [1924].

Rice, G. W. (ed.), *Christchurch Changing: An Illustrated History*, Christchurch, 1999.

Riley, J. T., *Reminiscences of Fifty Years in Western Australia*, Perth, 1903.

Rochford, J., *Adventures of a Surveyor in New Zealand and Australian Gold Diggings*, London, 1853.

Rolt, L. T. C., *Victorian Engineering*, London, 1970.

Scholefield, G. H., *Dictionary of New Zealand Biography*, Wellington, 1990.

Shaw, George Bernard, *Our Theatres in the Nineties*, vol. 3, London, 1954.

Sinclair, K., *Oxford Illustrated History of New Zealand*, Auckland, 1997.

Spillman, Ken, *A Rich Endowment: Government and Mining in Western Australia 1820–1994*, Nedlands, 1993.

Stannage, C. T., *The People of Perth*, Perth, 1979.

Stevens, J. W. B., *A History of the Harbour*, Fremantle, 1929.

Stewart, William Downie, *William Rolleston: A New Zealand Statesman*, Wellington, 1940.

Straub, Hans, *History of Civil Engineering*, London, 1952.

Tauman, Merab, *The Chief: C. Y. O'Connor*, Nedlands, 1978.

Trevelyan, George, *Life & Letters of Macaulay*, vol. 2, London, 1876.

Trinder, B., *The Making of the Industrial Landscape*, London, 1982.

Trollope, A., *South Australia, Western Australia and N.Z.*, London, 1884.

Truthful Thomas, *Through the Spyglass*, Perth, 1905.

Tull, Malcolm, *A Community Enterprise: The History of the Port of Fremantle, 1897–1997*, Newfoundland, 1997.

Twopenny, R. E. N., *Town Life in Australia*, London, 1883.

Vaughan, W. E. (ed.), *A New History of Ireland*, vol. V, *1801–1870*, Oxford, 1989.

Webb, M. J., 'The death of a hero: The strange suicide of Chas Y. O'Connor', *Early Days*, Royal Western Australian Historical Society journal, vol. II, part 1, 1995, pp. 81–111.

Webb, Martyn & Audrey, *Golden Destiny: The Centenary History of Kalgoorlie-Boulder and the Eastern Goldfields of Western Australia*, Kalgoorlie-Boulder, 1993.

Weir, Joan, *Back Door to the Klondike: An Account of Viscount Algernon Yelverton's Adventures Gold Prospecting in 1898*, Ontario, 1988.

West Coast Historical Museum, *Larrikins' Lode: Episodes in the History of the Kumara Goldrush*, Hokitika, 1998.

Whittington, Vera, *Gold and Typhoid: Two Fevers*, Nedlands, 1988.

Wilde, William R., *The Beauties and Antiquities of the Boyne*, facsimile edn Cork, 1978.

Woodham-Smith, Cecil, *The Great Hunger: Ireland 1845–49*, London, 1964 [1962].

Young, G. M., *Victorian England: Portrait of An Age*, London, 1986 [1936].

Unpublished Works

Coghlan, Maureen F., 'Monuments in Stone', thesis, Claremont Teachers College, 1958, BLP.

Crowley, F. K., 'Forrest the Politician 1891–1918', manuscript, BLP.

Greenwood, Elizabeth, 'Suicide in Adults', research paper, 1966.

Hunt, H. E., 'Chronological Examination of Design and Construction of the Coolgardie Pipeline', and other unpublished papers prepared for the National Trust of Western Australia, 2000.

Manford, T., 'A History of Rail Transport Policy in Western Australia 1870–1911', PhD thesis, UWA, 1976.

McKenzie, J., 'C. Y. O'Connor and Some Aspects of Public Works in Western Australia 1891–1902', BLP.

Milner, D. P., 'The Birth of the Public Works Department and Major Works', thesis, Maylands Teachers College, 1961, BLP.

Index

Notes

O'Connor's engineering works appear under their general title alphabetically. All references in the text to the pipeline scheme, even where the more popular or shortened form is used, can be found under the original title, Coolgardie Goldfields Water Supply Scheme. Photographs are listed in bold type.

Abbreviations

C. Y. O'Connor: CYO; New Zealand: NZ; Western Australia: WA

Printed in Australia
AUOC01n0638260716

277689AU00010B/11/P